ADDISON WESLEY   WESTERN EDITION

Math Makes Sense

5

Author Team

Peggy Morrow · Ralph Connelly

Ray Appel · Daryl M.J. Chichak

Bryn Keyes · Jason Johnston

Don Jones · Michael Davis

Steve Thomas · Jeananne Thomas

Angela D'Alessandro · Maggie Martin Connell

Trevor Brown · Sharon Jeroski

PEARSON
Addison Wesley

Publishing Team
Enid Haley
Lesley Haynes
Lynne Gulliver
Susan Lishman
Cecilia Chan
Keltie Thomas
Amanda Allan
Marg Bukta
Stephanie Cox
Judy Wilson
John Burnett
Jim Mennie
Nicole Argyropoulos

Publishers
Claire Burnett
Reid McAlpine

Elementary Math Team Leader
Anne-Marie Scullion

Product Manager
Nashaant Sanghavi

Photo Research
Karen Hunter

Design
Word & Image Design Studio Inc.

Copyright © 2005 Pearson Education Canada Inc.

All Rights Reserved. This publication is protected by copyright, and permission should be obtained from the publisher prior to any prohibited reproduction, storage in a retrieval system, or transmission in any form or by any means, electronic, mechanical, photocopying, recording, or likewise. For information regarding permission, write to the Permissions Department.

ISBN 0-321-24306-4

Printed and bound in the United States.

1 2 3 4 5 -- VHP -- 08 07 06 05 04

The information and activities presented in this book have been carefully edited and reviewed. However, the publisher shall not be liable for any damages resulting, in whole or in part, from the reader's use of this material.

Brand names that appear in photographs of products in this textbook are intended to provide students with a sense of the real-world applications of mathematics and are in no way intended to endorse specific products.

The publisher has taken every care to meet or exceed industry specifications for the manufacturing of textbooks. The spine and the endpapers of this sewn book have been reinforced with special fabric for extra binding strength. The cover is a premium, polymer-reinforced material designed to provide long life and withstand rugged use. Mylar gloss lamination has been applied for further durability.

Program Consultants and Advisers

Program Consultants

Craig Featherstone
Mignonne Wood
Maggie Martin Connell
Trevor Brown

Assessment Consultant
Sharon Jeroski

Elementary Mathematics Adviser
John A. Van de Walle

Program Advisers

Pearson Education thanks its Program Advisers, who helped shape the vision for *Addison Wesley Mathematics Makes Sense* through discussions and reviews of prototype materials and manuscript.

Anthony Azzopardi
Sandra Ball
Victoria Barlow
Lorraine Baron
Bob Belcher
Judy Blake
Steve Cairns
Christina Chambers
Daryl M.J. Chichak
Lynda Colgan
Marg Craig
Jennifer Gardner
Florence Glanfield
Linden Gray
Pamela Hagen
Dennis Hamaguchi
Angie Harding
Andrea Helmer
Peggy Hill

Auriana Kowalchuk
Gordon Li
Werner Liedtke
Jodi Mackie
Lois Marchand
Betty Milne
Cathy Molinski
Cynthia Pratt Nicolson
Bill Nimigon
Stephen Parks
Eileen Phillips
Carole Saundry
Evelyn Sawicki
Leyton Schnellert
Shannon Sharp
Michelle Skene
Lynn Strangway
Mignonne Wood

Program Reviewers

Field Testers

Pearson Education would like to thank the teachers and students who field-tested *Addison Wesley Math Makes Sense 5* prior to publication. Their feedback and constructive recommendations have been most valuable in helping us to develop a quality mathematics program.

Aboriginal Content Reviewers

Early Childhood and School Services Division
Department of Education, Culture, and Employment
Government of Northwest Territories:

Steven Daniel, Coordinator, Mathematics, Science, and Secondary Education
Liz Fowler, Coordinator, Culture Based Education
Margaret Erasmus, Coordinator, Aboriginal Languages

Grade 5 Reviewers

Mike Adam
Lambton-Kent District School Board, ON

Nora L. Alexander
Durham District School Board, ON

Lori Bryden
Catholic District School Board of Eastern Ontario, ON

Elvira Castaneda
Calgary Separate School District, AB

Sandra Clark-Chaisson
Peterborough Victoria Northumberland and Clarington Catholic District School Board, ON

Debra Jackson Dennis
Ottawa-Carleton District School Board, ON

Linda Edwards
Toronto District School Board, ON

Norma Fraser
School District 83 (North Okanagan/Shuswap), BC

Lisa Graham
Peel District School Board, ON

Jennifer Healey
Ottawa-Carleton District School Board, ON

Heidi Heaver
Halton District School Board, ON

Pamela Howe
Peel District School Board, ON

Karen L. Kent
District School Board Ontario North East, ON

Denis Levesque
Halton District School Board, ON

Dale MacCormack
Peel District School Board, ON

Becky Matthews
Victoria, BC

Stephen Parks
Fredericton, NB

Sergio Pascucci
Peel District School Board, ON

Lesley Risinger
Peel District School Board, ON

Michelle Rodriguez
Peterborough Victoria Northumberland and Clarington Catholic District School Board, ON

Manuel Salvati
Durham District School Board, ON

Kelvin Todd
Durham District School Board, ON

Table of Contents

Cross Strand Investigation—Building Castles 2

UNIT 1 Number Patterns ....

Launch	Charity Fundraising	4
Lesson 1	Number Patterns and Pattern Rules	6
Lesson 2	Creating Number Patterns	9
Lesson 3	Modelling Patterns	12
Lesson 4	Using Patterns to Solve Problems	16
Lesson 5	Strategies Toolkit	20
Unit Review	Show What You Know	22
Unit Problem	Charity Fundraising	24

UNIT 2 Whole Numbers ....

Launch	On the Dairy Farm	26
Lesson 1	Representing, Comparing, and Ordering Numbers	28
Lesson 2	Prime and Composite Numbers	31
Lesson 3	Using Mental Math to Add	34
Lesson 4	Adding 3- and 4- Digit Numbers	37
Lesson 5	Using Mental Math to Subtract	40
Lesson 6	Subtracting with 4-Digit Numbers	43
Lesson 7	Multiplication and Division Facts to 144	46
World of Work	Banquet Coordinator	49
Game	Multiplication Tic-Tac-Toe	50
Lesson 8	Multiplying with Multiples of 10	51
Lesson 9	Using Mental Math to Multiply	54
Lesson 10	Solving Problems by Estimating	57
Lesson 11	Multiplying Whole Numbers	60
Lesson 12	Dividing Whole Numbers	64
Lesson 13	Solving Problems	68
Game	Less Is More	71
Lesson 14	Strategies Toolkit	72
Unit Review	Show What You Know	74
Unit Problem	On the Dairy Farm	76

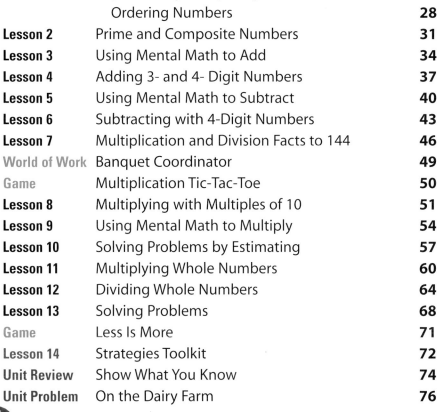

v

UNIT 3 Geometry

Launch	Bridges	78
Lesson 1	Identifying and Naming Polygons	80
Lesson 2	Constructing Triangles	84
Lesson 3	Combining Figures	87
Lesson 4	What Makes a Figure?	91
Lesson 5	Drawing Solids	96
Lesson 6	Planes of Symmetry	99
Lesson 7	Strategies Toolkit	102
Unit Review	Show What You Know	104
Unit Problem	Bridges	106

Cross Strand Investigation—Triangle, Triangle, Triangle   108

UNIT 4 Decimals

Launch	Coins Up Close	110
Lesson 1	Tenths and Hundredths	112
Lesson 2	Equivalent Decimals	116
Lesson 3	Comparing and Ordering Decimals	118
Lesson 4	Rounding Decimals	122
Lesson 5	Estimating Sums and Differences	125
Lesson 6	Adding Decimals	128
Game	Make Two!	131
Game	Spinning Decimals	132
Lesson 7	Subtracting Decimals	133
Lesson 8	Multiplying Decimals by 10 and 100	137
Lesson 9	Dividing Decimals by 10	141
Lesson 10	Strategies Toolkit	144
Unit Review	Show What You Know	146
Unit Problem	Coins Up Close	148

Cumulative Review   150

UNIT 5 Data Analysis

Launch	In the Lab	152
Lesson 1	Interpreting Data	154
Lesson 2	Frequency Tables and Line Plots	158
Technology	Creating Spreadsheets Using *AppleWorks*	162

	Lesson 3	Drawing Bar Graphs	164
	Technology	Drawing Bar Graphs Using *AppleWorks*	167
	Lesson 4	Broken-Line Graphs	169
	Technology	Drawing Broken-Line Graphs Using *AppleWorks*	173
	Lesson 5	Collecting Data	175
	Lesson 6	Making Inferences from Data	178
	Lesson 7	Strategies Toolkit	182
	Unit Review	Show What You Know	184
	Unit Problem	In the Lab	186

UNIT 6 Measurement

Launch	All Aboard!	188
Lesson 1	Measuring Time	190
Lesson 2	The 24-Hour Clock	193
Lesson 3	Exploring Time and Distance	197
Lesson 4	Strategies Toolkit	200
Lesson 5	Estimating and Counting Money	202
Lesson 6	Making Change	205
Lesson 7	Capacity	208
Lesson 8	Volume	210
Lesson 9	Relating Capacity and Volume	213
Lesson 10	Measuring Mass	216
Lesson 11	Exploring Large Masses	219
Unit Review	Show What You Know	222
Unit Problem	All Aboard!	224

UNIT 7 Transformational Geometry

Launch	Geometry in Art	226
Lesson 1	Translations	228
Lesson 2	Rotations	232
Lesson 3	Reflections	236
Lesson 4	Line Symmetry	240
Lesson 5	Exploring Tiling	243
Lesson 6	Strategies Toolkit	246
Lesson 7	Coordinate Grids	248
Unit Review	Show What You Know	252
Unit Problem	Geometry in Art	254

Cross Strand Investigation—Rep-Tiles 256

UNIT 8 — Fractions and Decimals

Launch	In the Garden	258
Lesson 1	Equivalent Fractions	260
Lesson 2	Fractions and Mixed Numbers	264
Lesson 3	Comparing and Ordering Fractions	267
Game	Order Up!	271
Lesson 4	Relating Fractions to Decimals	272
Lesson 5	Fraction and Decimal Benchmarks	276
Lesson 6	Relating Fractions to Division	279
Technology	Fractions and Decimals on a Calculator	282
Game	Fractions In-Between	283
Lesson 7	Estimating Products and Quotients	284
Lesson 8	Multiplying Decimals with Tenths	287
Lesson 9	Multiplying Decimals with Hundredths	291
Lesson 10	Strategies Toolkit	294
Lesson 11	Dividing Decimals with Tenths	296
Lesson 12	Dividing Decimals with Hundredths	299
Unit Review	Show What You Know	302
Unit Problem	In the Garden	304

Cumulative Review — 306

UNIT 9 — Length, Perimeter, and Area

Launch	At the Zoo	308
Lesson 1	Measuring Linear Dimensions	310
Lesson 2	Relating Units of Measure	313
Lesson 3	Using Non-Standard Units to Estimate Lengths	316
Lesson 4	Measuring Distance Around a Circular Object	318
Lesson 5	Using Grids to Find Perimeter and Area	321
Lesson 6	Measuring to Find Perimeter	325
Lesson 7	Perimeter and Area	329
Lesson 8	Finding the Area of an Irregular Polygon	333
Lesson 9	Estimating Area	337
Lesson 10	Strategies Toolkit	340
Unit Review	Show What You Know	342
Unit Problem	At the Zoo	344

UNIT 10 — Patterns in Number and Geometry

Launch	Squares Everywhere	346
Lesson 1	Patterns in Multiplication	348
Lesson 2	Graphing Patterns	352
Lesson 3	Another Number Pattern	356
World of Work	Choreographer	359
Lesson 4	Strategies Toolkit	360
Lesson 5	Tiling Patterns	362
Technology	Using a Computer to Create Tiling Patterns	365
Unit Review	Show What You Know	368
Unit Problem	Squares Everywhere	370

UNIT 11 — Probability

Launch	At the Pet Store!	372
Lesson 1	The Likelihood of Events	374
Lesson 2	Conducting Experiments	377
Lesson 3	Probability and Fractions	380
Lesson 4	Probability in Games	383
Lesson 5	Strategies Toolkit	386
Unit Review	Show What You Know	388
Unit Problem	At the Pet Store!	390

Cross Strand Investigation—The Domino Effect	392
Cumulative Review	394
Illustrated Glossary	398
Index	407
Acknowledgments	410

ix

Welcome to Addison Wesley Math Makes Sense 5

Math helps you to understand what you see and do every day.

You will use this book to learn about the math around you. Here's how.

In each Unit:

- A scene from the world around you reminds you of some of the math you already know.

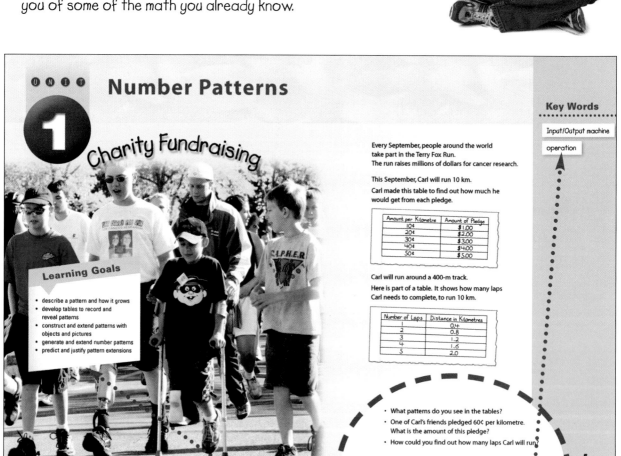

Find out what you will learn in the **Learning Goals** and important **Key Words**.

In each Lesson:

You **Explore** an idea or problem, usually with a partner. You often use materials.

Then you **Show and Share** your results with other students.

Practice questions help you to use and remember the math.

reminds you to use pictures, words, or numbers in your answers.

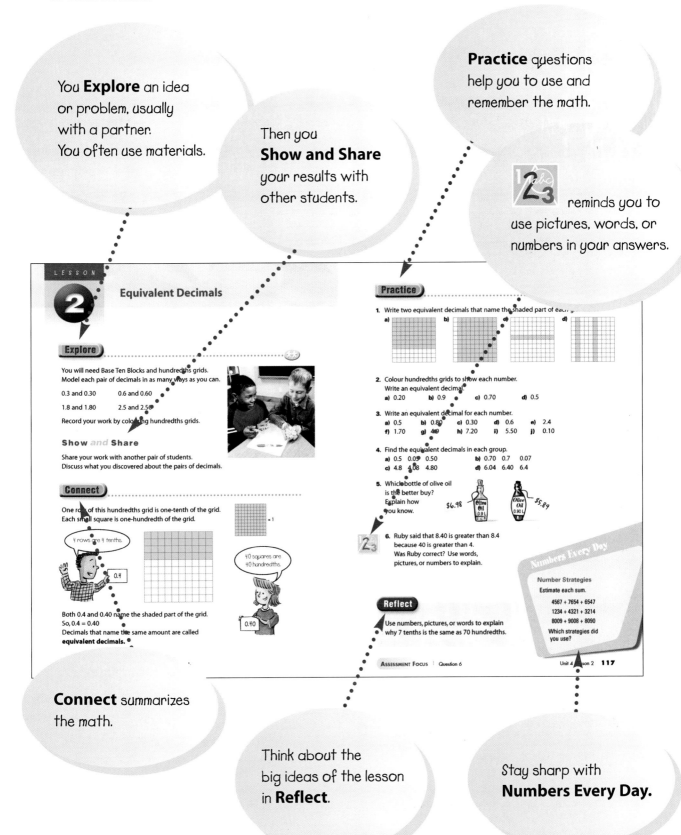

Connect summarizes the math.

Think about the big ideas of the lesson in **Reflect**.

Stay sharp with **Numbers Every Day.**

- Learn about strategies to help you solve problems in each **Strategies Toolkit** lesson.

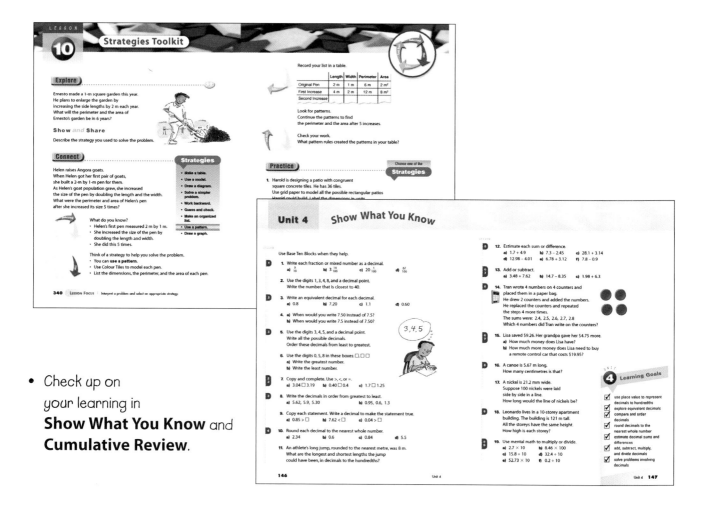

- Check up on your learning in **Show What You Know** and **Cumulative Review**.

- The **Unit Problem** returns to the opening scene. It presents a problem to solve or a project to do using the math of the unit.

xii

Explore some interesting math when you do the **Investigations**.

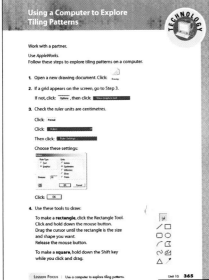

Use **Technology**.
Follow the step-by-step instructions for using a calculator or computer to do math.

Look for and .

You will see **World of Work** and **Games** pages.

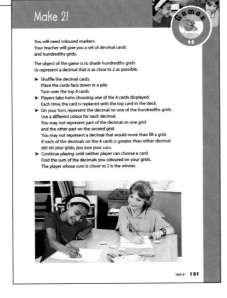

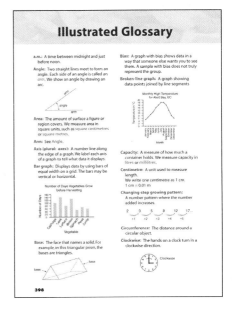

The **Glossary** is an illustrated dictionary of important math words.

xiii

Cross Strand Investigation

Building Castles

You will need Pattern Blocks.
Be sure you have squares and triangles.

Part 1

Look at this pattern.

 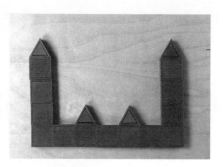

Frame 1 Frame 2 Frame 3

How many squares are in each frame?
How many triangles are in each frame?

Each block has a side length of 1 unit.
What is the perimeter of each frame?

Record the frame number, number of squares, number of triangles, and perimeter in a table.

Part 2

➤ Build Frame 4.
How many squares and triangles did you use?
What is the perimeter?
Record the data in your table.

➤ How many squares and triangles will you need to build Frame 5?
How did you find out?
Build Frame 5 to check your prediction.

➤ Predict the number of squares and triangles needed to build Frame 10.
How did you make your prediction?

➤ Write each pattern rule:
  • the numbers of squares in the frames
  • the numbers of triangles in the frames
  • the perimeters of the frames

Display Your Work

Record your work.
Describe the patterns you discovered.

Take It Further

Choose three different Pattern Blocks.
Build your own pattern.
Sketch the first 4 frames.
What number patterns can you find?

UNIT 1
Number Patterns
Charity Fundraising

Learning Goals

- describe a pattern and how it grows
- develop tables to record and reveal patterns
- construct and extend patterns with objects and pictures
- generate and extend number patterns
- predict and justify pattern extensions

Key Words

Input/Output machine

operation

Every September, people around the world take part in the Terry Fox Run.
The run raises millions of dollars for cancer research.

This September, Carl will run 10 km.

Carl made this table to find out how much he would get from each pledge.

Amount per Kilometre	Amount of Pledge
10¢	$1.00
20¢	$2.00
30¢	$3.00
40¢	$4.00
50¢	$5.00

Carl will run around a 400-m track.

Here is part of a table. It shows how many laps Carl needs to complete, to run 10 km.

Number of Laps	Distance in Kilometres
1	0.4
2	0.8
3	1.2
4	1.6
5	2.0

- What patterns do you see in the tables?
- One of Carl's friends pledged 60¢ per kilometre. What is the amount of this pledge?
- How could you find out how many laps Carl will run?

LESSON 1

Number Patterns and Pattern Rules

Is each pattern a growing pattern, a shrinking pattern, or a repeating pattern?
- 9, 14, 19, 24, 29, …
- 8, 5, 1, 8, 5, 1, …
- 48, 46, 44, 42, 40, …
- 4, 8, 16, 32, 64, …

What is each pattern rule?

Explore

For each number pattern:
- 3, 4, 6, 9, 13, …
- 3, 4, 6, 7, 9, …
- 1, 4, 3, 6, 5, 8, …
- 1, 2, 5, 10, 17, 26, …

➤ Identify the pattern rule.
   Write the next 5 terms.

➤ Make up a similar pattern.
   Trade patterns with a classmate.
   Write the rule for each of your classmate's patterns.

Show and Share

Share your patterns with other classmates.
How do you know each pattern rule is correct?
For any pattern, did you find more than one rule?
Explain.

Connect

 Here is a number pattern.

5 6 8 11 15 …
  +1 +2 +3 +4

The pattern rule is:

Start at 5. Add 1.
Increase the number you add by 1 each time.

To get the next 5 terms, continue to increase the number you add by 1 each time.
5, 6, 8, 11, 15, 20, 26, 33, 41, 50, …

 Here is another number pattern.

2 5 9 12 16 …
  +3 +4 +3 +4

The pattern rule is:

Start at 2. Alternately add 3, then add 4.

To get the next 5 terms, continue to add 3, then add 4.
2, 5, 9, 12, 16, 19, 23, 26, 30, 33, …

 Here is another number pattern.

10 6 11 7 12 …
  −4 +5 −4 +5

The pattern rule is:

Start at 10. Alternately subtract 4, then add 5.

To get the next 5 terms, continue to subtract 4, then add 5.
10, 6, 11, 7, 12, 8, 13, 9, 14, 10, …

> To identify the pattern rule, I find the difference between pairs of consecutive numbers in the pattern.

Numbers Every Day

Number Strategies

Estimate each difference.

357 − 85
423 − 176
5652 − 609

Which strategies did you use?

Unit 1 Lesson 1   **7**

Practice

Use a calculator when it helps.

1. Write the first 5 terms of each pattern.
   a) Start at 3. Add 9 each time.
   b) Start at 5. Add 2. Increase the number you add by 2 each time.
   c) Start at 7. Alternately add 3, then subtract 1.

2. Write the next 4 terms in each pattern. Write each pattern rule.
   a) 1, 2, 4, 5, 7, 8, …
   b) 2, 4, 3, 5, 4, 6, 5, …
   c) 98, 85, 87, 74, 76, …
   d) 1, 10, 7, 70, 67, 670, …

3. Find each missing term. Write the pattern rule.
   a) 3, 23, 13, 33, ☐, 43, 33, …
   b) 99, 98, 198, 197, ☐, 296, 396, …
   c) 2, 22, 12, 132, 122, 1342, ☐, …

4. What is the 7th term of this pattern?
   Start at 200. Subtract 8 each time.
   How could you find the 7th term without writing the first 6 terms?

5. Find each missing term. Write the pattern rule.
   a) 74, 148, 222, ☐, 370, …
   b) 100, 198, 295, 391, ☐, 580, …
   c) 1122, 1112, 1101, 1091, 1080, ☐, 1059, …

6. What is the 10th term of this pattern?
   Start at 13. Alternately subtract 4, then add 5.

7. The first 2 terms of a pattern are 6, 12, ….
   How many different patterns can you write with these 2 terms?
   For each pattern, list the first 6 terms and write the pattern rule.
   Show your work.

Reflect

How do you find the pattern rule for a number pattern?
Use an example to explain.

LESSON 2

Creating Number Patterns

Look at this **Input/Output machine**.
Any number that is put into this
machine is multiplied by 5.
If you input 5, the output is 25.
If you input 9, the output is 45.

Explore

➤ Draw your own Input/Output machine.
   Choose a number to go inside your machine.
   Choose an operation.
   Use your machine to create a number pattern.
➤ Copy and complete this table for your pattern.
   Write the pattern rule for the output numbers.

An **operation** is add, subtract, multiply, or divide.

Input	Output
1	
2	
3	

Show and Share

Share your machine and table
with another pair of classmates.
Use your classmates' machine to extend their number pattern.

Connect

We can use an Input/Output machine
to make a growing pattern.
Each input is added to 8 to get the output.
When we input 1, the output is 9.
When we input 2, the output is 10.

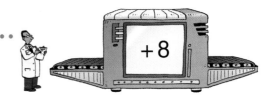

Input	Output
1	9
2	10
3	11
4	12

The pattern rule for the output is:

Start at 9. Add 1 each time.

LESSON FOCUS | Use an Input/Output machine to create number patterns.

Use a calculator when it helps.

1. For each Input/Output machine:
   - Choose 5 numbers to input.
   - Find each output number.
   - Copy and complete the table.

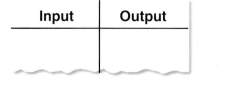

Input	Output

   a)    b)

2. Copy and complete the table for this Input/Output machine.

   Input → [− 7] → Output

Input	Output
20	
19	
18	
17	
16	

3. a) Copy and complete the table for this Input/Output machine.

   Input → [÷ 6] → Output

Input	Output
36	
42	
48	
54	
60	
66	
72	

   Number Sense

   When an Input/Output machine uses multiplication, and the input numbers are whole numbers, the output numbers are multiples of the number in the machine.

   b) What is the pattern rule for the input numbers? The output numbers?

4. Each table shows the Input/Output from a machine.
   - Identify the number and the operation in the machine.
   - Continue the patterns.
     Write the next 4 input and output numbers for each table.

a)

Input	Output
2	20
4	40
6	60
8	80
10	100

b)

Input	Output
500	485
450	435
400	385
350	335
300	285

5. For each table:
   - Write the pattern rules for the output numbers and the input numbers.
   - The patterns continue. Write the next 4 input and output numbers.

a)

Input	Output
110	101
99	90
88	79
77	68
66	57

b)

Input	Output
84	12
91	13
98	14
105	15
112	16
119	17

6. Draw an Input/Output machine.
   Choose a number and an operation.
   Use multiples of any number as the input numbers.
   Find the output numbers.
   Make a table to show your results.
   Write the pattern rules for the input numbers and the output numbers.

Reflect

When you see an Input/Output table, how can you tell what the operation is in the machine? Use words, pictures, or numbers to explain.

Numbers Every Day

Mental Math

Add.

156 + 302

411 + 298

2356 + 1009

Which strategies did you use?

LESSON

Modelling Patterns

Explore

You will need a geoboard, geobands, and dot paper.

➤ Use the geoboard to make a rectangle with length 2 units and width 1 unit. Count and record the number of pegs on the perimeter of the rectangle.
➤ Make a rectangle with length 3 units and width 2 units. Count and record the number of pegs on the perimeter.
➤ Continue to make rectangles with length 1 unit greater than the width. Record the length, the width, and the number of pegs each time.

Draw each rectangle on dot paper.

Rectangle	Length	Width	Number of Pegs on Perimeter
1	2	1	6
2	3	2	

- How many pegs will be on the perimeter of the 10th rectangle? The 20th rectangle?
- Will the perimeter of any rectangle have 42 pegs? 44 pegs? 46 pegs? How do you know?

Show and Share

Share your results with another pair of classmates.
What patterns do you see in the table?
How did you use these patterns to solve the problems?

12   LESSON FOCUS | Model patterns and make predictions.

Connect

Here is a pattern of similar triangles drawn on dot paper. Each triangle has 2 equal sides.

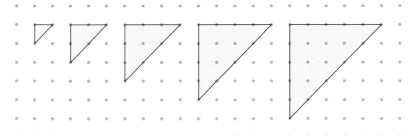

Similar triangles have the same shape.

This pattern continues.
➤ Find the number of dots on the perimeter of the 10th triangle and the 20th triangle.

Make a table.

Triangle	Number of Dots on Perimeter
1	3
2	6
3	9
4	12
5	15

These numbers are multiples of 3.

 One pattern rule for the number of dots on the perimeter is:

Start at 3. Add 3 each time.

 Another pattern rule for the number of dots is:

The triangle number multiplied by 3

The 10th triangle will have 10 × 3, or 30 dots on its perimeter.
The 20th triangle will have 20 × 3, or 60 dots on its perimeter.

➤ Will any triangle have a perimeter of 50 dots? 51 dots?
Use the pattern to find out.

The number of dots on any perimeter is a multiple of 3.
50 is not a multiple of 3, so no triangle has 50 dots.
51 is a multiple of 3 because 17 × 3 = 51.
So, the 17th triangle has 51 dots.

Practice

1. Regular pentagons are combined to make new figures. Each pentagon touches no more than 2 other pentagons.

   The side length of each pentagon is 1 unit.
   The perimeter of each figure is recorded in a table.

Number of Pentagons	Perimeter (units)
1	5

   a) Copy and complete the table for the first 4 figures.
   b) Write the pattern rule for the perimeters.
   c) Use the pattern to predict the perimeter of the figure with 10 pentagons. With 20 pentagons.

2. Here is a pattern of figures made with Colour Tiles.

   Figure 1   Figure 2   Figure 3   Figure 4

   The pattern continues.
   a) Draw the next two figures on grid paper.
   b) Copy and complete the table for the first 6 figures.

Figure	Number of Green Tiles	Number of Yellow Tiles
1	2	10

   c) Write the pattern rule for the number of green tiles. The number of yellow tiles.
   d) How many green tiles will be in the 15th figure?
   e) How many yellow tiles will be in the 20th figure?
   f) Will any figure have 31 green tiles? 41 yellow tiles? Explain.

3. Here is a pattern of Snap Cubes.

  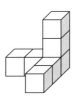 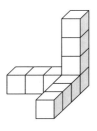

   Object 1   Object 2   Object 3   Object 4

   The pattern continues. Use Snap Cubes.
   a) Make the next two objects.
   b) Copy and complete this table for the first 6 objects.

Object	Number of Cubes
1	1

   c) How does the pattern grow? Write the pattern rule for the number of cubes.
   d) How many cubes will there be in the 10th object? How do you know?
   e) Will any object have 50 cubes? 51 cubes? How do you know?

4. Dreamy Ice-Cream Company sells a single-scoop sundae for $1.25. Each additional scoop costs $1.25. There is a fixed price of 50¢ for extra topping on all sundaes.
   a) Make a table to show the prices of the first 4 sizes of sundaes with extra topping.
   b) Write a pattern rule for any patterns you see.
   c) What is the price of a 6-scoop sundae with extra topping? Explain. Show your work.

Reflect

Use a question from this lesson. How can patterns in a table help you solve a problem, without extending the table? Use words, pictures, or numbers to explain.

Numbers Every Day

Number Strategies

Order the decimals in each set from least to greatest.

2.3, 1.3, 10.1

1.98, 2.01, 1.89

0.35, 0.23, 2.30

ASSESSMENT FOCUS | Question 4

LESSON 4

Using Patterns to Solve Problems

Sam baby-sits to make money.
Sam charges $6 for each hour he works.

➤ How much does Sam earn when he works
2 hours? 3 hours? 4 hours? 5 hours?
Show your results in a table.

Time Worked (hours)	Money Earned ($)

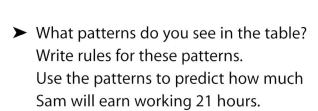

➤ What patterns do you see in the table?
Write rules for these patterns.
Use the patterns to predict how much
Sam will earn working 21 hours.

➤ Sam saves all the money he earns.
He needs $250 to buy a mountain bike.
How many hours does Sam need to work?

➤ Make up your own problem you can solve using this table.
Trade problems with a classmate.
Solve your classmate's problem.

Show and Share

Share your answers with a classmate.
Did you solve the problems the same way? Explain.

LESSON FOCUS | Pose and solve problems by applying a patterning strategy.

One puzzle book costs $17.

➤ How much does it cost to buy 2 books? 3 books? 4 books?

Make a table.
When you add 1 to the number of books,
you add $17 to the cost.

Two books cost $34.
Three books cost $51.
Four books cost $68.

Number of Books	Cost ($)
1	17
2	34
3	51
4	68

These numbers are multiples of 17.

➤ Use a pattern to predict the cost of 20 books.

 One pattern rule for the cost is:

Start at 17. Add 17 each time.

 Another pattern rule for the cost is:

The number of books multiplied by 17

To predict the cost of 20 books, multiply: 20 × 17 = 340
Twenty books cost $340.

➤ Extend the pattern to find how many books you can buy with $200.
Use a calculator.

Press: [6] [8] [+] [1] [7] [=] [=] [=] [=] [=] [=] ...

Record each number in the table.

Continue to press [=] until the number is greater than 200.

Eleven books cost $187.
Twelve books cost $204.
So, you can buy 11 books with $200.

Number of Books	Cost ($)
1	17
2	34
3	51
4	68
5	85
6	102
7	119
8	136
9	153
10	170
11	187
12	204

Unit 1 Lesson 4

Practice

1. The pattern in this table continues.

Number	Cost ($)
1	16
2	32
3	48
4	64
5	
6	
7	

Numbers Every Day

Mental Math

Multiply.

5×30
60×4
700×6
9×8000

Which strategies did you use?

   a) Copy and complete the table.
   b) Write a pattern rule for the cost.
   c) Write a story problem you could solve using this table.
      Solve your problem.

2. Hilary has a paper route. Each week she collects $31.
   a) How much money has Hilary collected at the end of 1 week? 2 weeks?
   b) Make a table to show the amounts for the first 8 weeks.
   c) How much will Hilary collect in total in 3 weeks?
   d) How much will Hilary collect in 1 year?
   e) Write a problem you could solve using the table in part a.
      Solve your problem.

3. The sunflower is the only single flower that grows as high as 3 m.
   It can grow 30 cm each week.
   In which week could a sunflower reach a height of 3 m? Explain.

4. Look at this figure.
   a) How many triangles are there with side length 1 unit? 2 units? 3 units?
   b) How many triangles are in this figure?

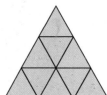

5. How many squares are in this figure?

6. Dave read 70 pages on Monday, 67 pages on Tuesday, and 64 pages on Wednesday.
   This pattern continued until Dave finished his book.
   a) Make a table to show how many pages were read each day.
   b) What was the total number of pages Dave read the first 7 days?
   c) Dave finished his book on the day he read 31 pages.
      How many pages are in the book?
   Show your work.

7. Greenholme Elementary School has a car wash to raise money for the community.
   The students charge $4 to wash a car.
   a) Make a table to show how much money is collected when these numbers of cars are washed:
      10, 20, 30, 40, 50

The students spent $23 for soap and sponges. This amount is subtracted from the amount collected.
   b) How many cars have to be washed to raise $350?
   c) Write your own problem about this car wash. Solve the problem you posed.

Reflect

How can using patterns help you solve problems? Use an example from this lesson to explain.

At Home

What number patterns do you see at home? Look through magazines and newspapers. Write about the patterns you see.

ASSESSMENT FOCUS | Question 6   Unit 1 Lesson 4   **19**

LESSON 5: Strategies Toolkit

Explore

Abi made an Input/Output machine that uses two operations.

Here is a table for Abi's machine.

Find out what the machine does to each input number.

Input → → Output

Input	Output
1	5
2	7
3	9
4	11
5	13

Show and Share

Explain the strategy you used to solve the problem.

Connect

Ben made an Input/Output machine that uses two operations.
Here is a table for Ben's machine. What does Ben's machine do to each input number?

Input	Output
1	2
2	5
3	8
4	11
5	14

Strategies

- Make a table.
- Use a model.
- Draw a diagram.
- Solve a simpler problem.
- Work backward.
- Guess and check.
- Make an organized list.
- Use a pattern.
- Draw a graph.

What do you know?
- The machine uses two operations on an input number. The operations could be add, subtract, multiply, or divide.

Think of a strategy to help you solve the problem.
- You can **use a pattern**.
- Analyse the pattern in the *Output* column to find out what the machine does to each input number.

20 LESSON FOCUS | Interpret a problem and select an appropriate strategy.

The numbers in the Output column increase by 3. This suggests the pattern involves multiples of 3. Which two operations does Ben's machine use?

Input	Output
1	2
2	5
3	8
4	11
5	14

Find the pattern rule for the output numbers. Extend this pattern.
Use the operations in the machine to extend the pattern of the output numbers.
Check the numbers in both patterns match.

Practice

Choose one of the Strategies

1. Design an Input/Output machine that gives these results. How did you decide which operations to use?

Input	Output
1	9
2	14
3	19
4	24
5	29

2. Annie and her 6 friends exchange friendship bracelets. How many bracelets are needed so each person gets a bracelet from each other person?

Reflect

Choose one of the questions in this lesson. Explain how you used a pattern to solve it.

Unit 1 Lesson 5

Unit 1 Show What You Know

LESSON

1

1. Write the first 6 terms of each pattern.
   a) Start at 100. Subtract 9 each time.
   b) Start at 10. Alternately add 5, then multiply by 2.

2. Write the next 5 terms in each pattern.
   Write the pattern rule.
   a) 5, 8, 12, 15, 19, …
   b) 50, 48, 47, 45, 44, …
   c) 10, 12, 16, 22, 30, …

2

3. For this Input/Output machine:

   Input → + 11 → Output

Input	Output

   a) Choose 6 numbers to input.
   b) Find each output number.
   c) Copy and complete the table.

4. For each table:
   - Write the pattern rules for the input numbers and the output numbers.
   - The patterns continue.
     Write the next 5 input and output numbers.
   - What do you do to each input number to get the output number?

   a)
Input	Output
240	20
216	18
192	16
168	14
144	12

   b)
Input	Output
21	189
31	279
41	369
51	459
61	549

LESSON 3

5. Congruent hexagons are combined to make new figures. Each hexagon touches no more than 2 other hexagons.

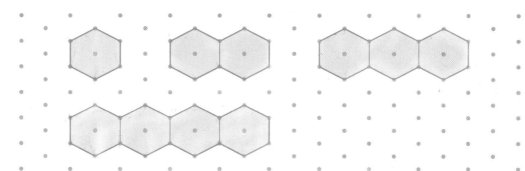

The side length of each hexagon is 1 unit.
The perimeter of each figure is recorded in a table.
The pattern continues.
Use Pattern Blocks when they help.

a) Copy and complete the table.
b) Write the pattern rule for the perimeters.
c) Use the pattern to predict the perimeter of the figure with 8 hexagons.
d) What is the perimeter of the figure with 15 hexagons?
e) Will a figure have a perimeter of 30 units? 40 units? Explain.

Number of Hexagons	Perimeter (units)
1	6
2	10
3	
4	
5	

6. A magazine costs $3.50.

a) What is the cost of 2 magazines? 3 magazines? 4 magazines? 5 magazines? 6 magazines? Show your answers in a table.
b) How much would 98 magazines cost?
c) How many magazines can you buy with $100?
d) Write your own problem about these magazines. Solve the problem.

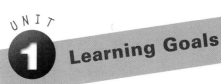

UNIT 1 Learning Goals

- ✓ describe a pattern and how it grows
- ✓ develop tables to record and reveal patterns
- ✓ construct and extend patterns with objects and pictures
- ✓ generate and extend number patterns
- ✓ predict and justify pattern extensions

Unit 1 **23**

Unit Problem: Charity Fundraising

Plan an event to raise money for charity.

Include:
- A description of the event
- How much you estimate the costs will be
- How much money you expect to raise
- Tables to show any patterns in the money you expect to raise
- A poster to promote your fundraising event

Check List

Your work should show

- ☑ a detailed plan of the event
- ☑ how you calculate the amount you expect to raise
- ☑ any tables and patterns you used
- ☑ correct math language

Bake Sale

Pies - $7
Cakes - $9
Muffins - $2
Cookies - $1

Bike-A-Thon

Reflect on the Unit

Write about some of the different patterns in the unit, and how you used them to solve problems.

UNIT 2

Whole Numbers

On the Dairy Farm

Learning Goals

- read and write numbers to 999 999
- estimate to solve problems
- compare and order numbers
- use place value to represent numbers
- find prime and composite numbers
- find factors and multiples
- estimate sums, differences, products, and quotients
- add, subtract, and multiply numbers mentally
- add and subtract 4-digit numbers
- multiply a 3-digit number by a 2-digit number
- divide a 3-digit number by a 1-digit number
- solve problems using whole numbers
- solve problems with more than one step

Key Words

prime number

composite number

compensation

dividend

divisor

quotient

compatible numbers

- Hay is one part of a dairy cow's diet.
  140 kg of hay feeds 2 cows for 2 weeks.
  About how much hay do 2 cows eat each week? Each day?
- The Colyns have 30 dairy cows on their farm.
  Each day, they collect 27 L of milk from 1 cow.
  Estimate the amount of milk produced by 30 cows.

LESSON 1

Representing, Comparing, and Ordering Numbers

Here are some ways to represent the number 4027.

Standard Form
4027

Words
four thousand twenty-seven

Expanded Form
4000 + 20 + 7

Base Ten Blocks

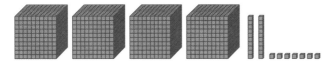

Place-Value Chart

Thousands	Hundreds	Tens	Ones
4	0	2	7

Explore

The heat of chili peppers is measured in Scoville units.
The greater the number of Scoville units, the hotter the pepper.

Danielle tested several types of peppers. Her results are shown in the table.

Which pepper is the hottest?
The mildest?
Order the peppers from mildest to hottest.

Pepper	Scoville Units
Cayenne	47 400
Jalapeno	4 960
Ancho	2 358
Tabasco	43 340
Habanero	103 050
Chipotle	91 880

Show and Share

Share your results with another pair of students. What strategies did you use to compare and order the numbers?

28 **LESSON FOCUS** | Represent, compare, and order numbers to 999 999.

The table shows the population of some Canadian cities in 2001.

City	Population
St. John's, NL	99 182
Halifax, NS	359 111
Quebec, QC	169 076
Winnipeg, MB	619 544
Edmonton, AB	666 104
Vancouver, BC	545 671

➤ Order the numbers from greatest to least.
Use a place-value chart.

City	Thousands			Units		
	Hundred Thousands	Ten Thousands	One Thousands	Hundreds	Tens	Ones
St. John's, NL		9	9	1	8	2
Halifax, NS	3	5	9	1	1	1
Quebec, QC	1	6	9	0	7	6
Winnipeg, MB	6	1	9	5	4	4
Edmonton, AB	6	6	6	1	0	4
Vancouver, BC	5	4	5	6	7	1

666 104 has the most hundred thousands and ten thousands.
It is the greatest number.

99 182 has no hundred thousands.
It is the least number.

When we write a number with more than 4 digits in standard form, we put a space between the hundreds digit and the thousands digit.

The order of the numbers from greatest to least is:
666 104, 619 544, 545 671, 359 111, 169 076, 99 182

➤ Edmonton has a population of 666 104.
We can write this in expanded form:
600 000 + 60 000 + 6000 + 100 + 4
and in words: six hundred sixty-six thousand one hundred four

Practice

1. Write each number in standard form.
   a) 600 000 + 20 000 + 50 + 7
   b) nine hundred fifty thousand six

2. Order the numbers from least to greatest.
   421 035   406 583   423 004   40 795

3. Write each number in question 2 in expanded form and in words.

4. Copy and complete. Replace each □ with >, < , or =.
   How did you decide which symbol to use?
   a) 35 937 □ 35 397
   b) 272 456 □ 227 456
   c) 456 123 □ 456 123
   d) 975 346 □ 985 346

5. Mariette wrote a 6-digit number. One digit was 0.
   The other digits were odd. No two digits were the same.
   The number was the greatest number
   she could write with these digits.
   What number did Mariette write? How do you know?

6. The digits in 234 567 are in order from least to greatest.
   Write 5 different 6-digit numbers with their digits in order
   from least to greatest.

7. Use the digits from 1 to 9. Use each digit only once.
   a) Make a 6-digit number as close to 100 000 as possible.
   b) Make a 6-digit number as close to 500 000 as possible.
   c) Which number did you get closest to?
      How do you know?

Reflect

Denis said 84 914 is greater than
311 902 because 8 is greater than 3.
Is Denis correct?
Use words and numbers to explain.

Numbers Every Day

Number Strategies

Copy and complete each pattern.
What is each pattern rule?

94, 86, 78, 70, □
48, 56, 64, 72, □
212, □, 196, 188
89, □, 105, 113

LESSON 2

Prime and Composite Numbers

Explore

You will need 12 Colour Tiles and grid paper.

➤ Arrange 2 tiles to make a rectangle.
How many different rectangles can you make?
Record each way on grid paper.

➤ Arrange 3 tiles to make a rectangle.
How many different rectangles can you make?
Record each way on grid paper.

➤ Repeat the activity with 4 tiles, then 5 tiles,
then 6 tiles, and so on, up to 12 tiles.

Remember that a 2 by 1 rectangle is the same as a 1 by 2 rectangle.

Show and Share

Compare rectangles with another pair of classmates.
For which numbers of tiles could you make only 1 rectangle?
2 rectangles? 3 rectangles?

LESSON FOCUS | Recognize, model, and describe prime and composite numbers.

31

Connect

➤ With 16 Colour Tiles, we can make 3 different rectangles.

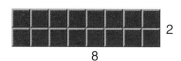

The side lengths of the rectangles are factors of 16.
We list the factors of 16: 1, 2, 4, 8, and 16
16 has five factors.
A number that has more than two factors is a **composite number**.

➤ With 17 Colour Tiles, we can make only 1 rectangle.

The side lengths of this rectangle are the factors of 17.
We list the factors of 17: 1, 17
17 has two factors.

A number that has exactly two factors is a prime number.

➤ Since the number 1 has only one factor,
1 is neither prime nor composite.

Practice

1. List the factors of each number.
   a) 2   b) 3   c) 4   d) 5   e) 6   f) 7
   g) 8   h) 9   i) 10   j) 11   k) 12

2. a) Which numbers from 2 to 12 are composite numbers?
      How do you know?
   b) Which numbers from 2 to 12 are prime numbers?
      How do you know?

3. Use Colour Tiles.
   Or draw rectangles on grid paper.
   Find all the factors of each number.
   a) 13   b) 14   c) 15   d) 18

   Which of these numbers are prime? Composite?
   How do you know?

4. Copy this table.

	Prime	Composite
Even		
Odd		

   a) Sort the numbers from 1 to 25 in the table.
   b) Which number does not belong in the table? Explain.

5. Look at a hundred chart.
   How can you use the chart to help you find
   all the prime numbers from 1 to 100?
   List these prime numbers.
   Show your work.

6. List all the prime numbers between 20 and 30.
   Explain how you know they are prime.

7. List all the composite numbers between 30 and 40.
   Explain how you know they are composite.

8. a) Are all odd numbers prime numbers? Explain.
   b) Are all even numbers composite? Explain.

Reflect

When you see a number, how can you tell
if it is a prime number or a composite number?
Use diagrams in your explanation.

Numbers Every Day

Number Strategies

Write each number in words.
What does the digit 5 represent
in each number?
- 45 302
- 90 215
- 58 760
- 11 542
- 30 051

ASSESSMENT FOCUS | Question 5

LESSON 3

Using Mental Math to Add

Explore

Workers at the Portage Bicycle Factory make mountain bikes and racing bikes. Last week, they made 2343 mountain bikes and 1998 racing bikes.

About how many bikes were made altogether? Find the total number of bikes that were made. Record your estimate and your sum. How close was your estimate to the sum?

Show and Share

Share your work with a classmate. Describe and compare the strategies you used to estimate and to find the total number of bikes.

Connect

➤ Estimate the sum: 3438 + 4279

Use front-end estimation.
Add the first digits of the numbers.
3438 + 4279 is about 3000 + 4000, or 7000.

For a better estimate, round each number to the nearest hundred.
3438 rounded to the nearest hundred is 3400.
4279 rounded to the nearest hundred is 4300.
3400 + 4300 = 7700

3438 + 4279 is about 7700.

I estimate when I don't need an exact answer. There are many ways to estimate.

34 **LESSON FOCUS** | Use different mental math strategies to add whole numbers.

➤ Use mental math to add: 2180 + 6432
Use **compensation**.
2180 + 20 = 2200
6432 − 20 = 6412
The sum of 2200 + 6412 is equal to the sum of 2180 + 6432.
2200 + 6412 = 8612

2180 + 6432 = 8612

To compensate, I took 20 away from 6432 and added it to 2180.

➤ Use mental math to add: 2625 + 432
Use the strategy of adding on.

2625 + 400 + 30 + 2

Start with the greater number, 2625.
There are 4 hundreds to add.
Count on by 100 four times:

2625   2725   2825   2925   3025
   +100   +100   +100   +100

There are 3 tens to add.
Count on by 10 three times:

3025   3035   3045   3055
   +10   +10   +10

Then add 2:
3055 + 2 = 3057

2625 + 432 = 3057

Practice

Use mental math.

1. Add. Use as many different strategies as you can.
   a) 6145 + 3007   b) 3654 + 372   c) 500 + 2150
   d) 1999 + 999   e) 4003 + 2968   f) 7741 + 685

2. Find the sums less than 10 000.
   a) 3099 + 5824   b) 6489 + 3201   c) 4673 + 6595
   d) 9997 + 8763   e) 5036 + 297   f) 9539 + 470

3. Natalie's school gave a concert last week.
   On Friday, 4560 people attended.
   On Saturday, 2995 people attended.
   How many people attended the concert in all?

4. Victoria shows her friends a number trick.
   She says:
   - Pick any 4-digit number.
     Don't tell me what it is.
   - Add 498 to the number.
   - Subtract 202 from the sum.
   - Add 204 to the difference.
   - Tell me what number you have now.
   - The number you picked is ☐.
   Victoria subtracts 500 from the number she is told.
   This always gives her the original number.

   Why does this trick work?
   Use words and numbers to explain.

5. A blueberry pie has 2285 calories.
   A cherry pie has 2465 calories.
   About how many calories are in both pies altogether?

6. Write an addition problem you can solve
   using mental math.
   Solve your problem.
   Which strategy did you use to solve
   your problem? Why?

Reflect

Describe 2 different ways you could
mentally add 4621 + 5393.

Numbers Every Day

Number Strategies

Round each number to the nearest ten, the nearest hundred, and the nearest thousand.
- 6493
- 7989
- 5091
- 8996
- 3003

LESSON 4

Adding 3- and 4-Digit Numbers

Explore

Last week, Sarah drank 1196 mL of juice.
Luke drank 937 mL.
How many millilitres of juice
did they drink altogether?

Show and Share

Share your solution with another pair of students.
How did you add the numbers?

Connect

Gemma ate 738 g of fruit one week.
In the same week, Devon ate 1452 g of fruit.
How many grams of fruit did they eat altogether?

Add: 738 + 1452

➤ Use Base Ten Blocks on a place-value mat to add.

- Trade 10 ones for 1 ten.

LESSON FOCUS | Use different strategies to add 3- and 4-digit numbers.

37

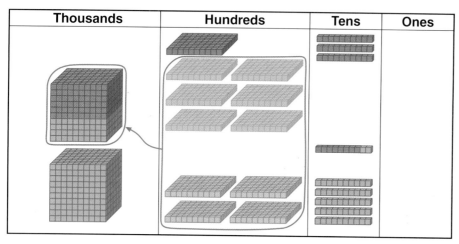

- Trade 10 hundreds for 1 thousand.

738 + 1452 = 2190

➤ Use expanded form to add.

$$
\begin{array}{rl}
738 & \rightarrow \quad\quad 700 + 30 + 8 \\
+\,1452 & \rightarrow \quad 1000 + 400 + 50 + 2 \\
\hline
& \quad\quad 1000 + 1100 + 80 + 10 \; = \; 2100 + 90 \; = \; 2190
\end{array}
$$

➤ Use place value to add.

Add the ones: 10 ones Regroup 10 ones as 1 ten 0 ones.	Add the tens: 9 tens	Add the hundreds: 11 hundreds Regroup 11 hundreds as 1 thousand 1 hundred.	Add the thousands: 2 thousands
$\overset{1}{7}38$ + 1452 ——— 0	$\overset{1}{7}38$ + 1452 ——— 90	$\overset{1\;1}{7}38$ + 1452 ——— 190	$\overset{1\;1}{8}38$ + 1452 ——— 2190

Together Gemma and Devon ate 2190 g of fruit last week.

Practice

1. Add.
   a) 8689
      + 798

   b) 6877
      + 2364

   c) 4062
      + 859

   d) 6953
      + 5241

2. Estimate. Then find each sum greater than 7000.
   a) 4176
      + 2457

   b) 3872
      + 5129

   c) 5839
      + 987

   d) 6518
      + 2828

3. Keshav collects stamps.
   He has 3845 Canadian stamps and
   2690 stamps from other countries.
   How many stamps does he have in all?

4. Two games were played in the semifinals
   of a hockey tournament.
   The attendance at one game was 9847.
   The attendance at the other game was 7788.
   How many people attended the semifinals?

5. Members of the school council raised $7000.
   They plan to buy sports equipment for $3596
   and library books for $3438.
   Did they raise enough money to make the purchases?
   How do you know?

6. Regional Recycling has a target of 2450 kg of aluminum.
   Suppose Fairfield delivers 1665 kg of aluminum,
   and Westdale delivers 795 kg of aluminum.
   Will Regional Recycling meet its goal? Explain.

7. Two 4-digit numbers have a sum of 9432.
   What might the numbers be? How do you know?

Reflect

Describe how you would use Base Ten Blocks
to add 981 + 3131.

Numbers Every Day

Mental Math

Add. Which strategies did you use?

443 + 268
114 + 508
611 + 294
407 + 342

ASSESSMENT FOCUS | Question 6

LESSON 5

Using Mental Math to Subtract

Explore

After the first snowfall of the year, 1184 people went snowboarding at Shred City Snowpark. After the second snowfall, 978 people went snowboarding.

➤ About how many more people went snowboarding after the first snowfall than after the second?

➤ How many more people went snowboarding after the first snowfall?

Record your estimate and the difference. How close was your estimate to the difference?

Show and Share

Share your work with a classmate. Describe and compare the strategies you used to estimate and to find the difference in the numbers of snowboarders.

Connect

➤ Estimate: 3818 − 2079

Round each number to the nearest hundred.
3818 rounded to the nearest hundred is 3800.
2079 rounded to the nearest hundred is 2100.
3800 − 2100 = 1700

3818 − 2079 is about 1700.

40   **LESSON FOCUS** | Use different mental math strategies to subtract whole numbers.

➤ Subtract: 2437 − 2298
Make a friendly number.

Add 2 to 2298 to make 2300.
Add 2 to 2437 to make 2439.

Write 2437 − 2298 as:
2439 − 2300 = 139

2437 − 2298 = 139

If I add 2 to each number, then subtract, the answer is the same as simply subtracting the two numbers.

➤ Use mental math to subtract: 5791 − 3486

Think of 3486 in expanded form:
3000 + 400 + 80 + 6
Subtract each number in turn.

5791 − 3000 = 2791
2791 − 400 = 2391
2391 − 80 = 2311
2311 − 6 = 2305

5791 − 3486 = 2305

Practice

Use mental math.

1. Subtract. Which strategy did you use each time?
   a) 7438 − 6002
   b) 5002 − 2797
   c) 4555 − 1998
   d) 4261 − 2256
   e) 6848 − 954
   f) 3011 − 427

2. Estimate.
   Then subtract the numbers for which the difference will be less than 4000.
   a) 5437 − 1601
   b) 5432 − 370
   c) 7715 − 2898
   d) 4060 − 256
   e) 6227 − 2419
   f) 8008 − 5050

3. George's car has a mass of 1262 kg.
   Sue's van has a mass of 1957 kg.
   Which vehicle has greater mass?
   How much greater?

Math Link

Your World

Jeanne Louise Calment of France is the oldest woman ever.
She lived from 1875 to 1997.
How many years did she live?

4. Cole has $1307 in his savings account.
   Emma has $988 in hers.
   How much more money does Cole have than Emma?

5. Last Saturday and Sunday, 3741 people visited the cultural centre.
   On Saturday, 2802 people visited the centre.
   How many people visited the centre on Sunday?

6. Write a subtraction problem you can solve using mental math.
   Solve the problem.
   Describe the strategy you used to solve the problem.
   Explain your choice.

7. The answer to a subtraction problem is 2550.
   Use mental math to find 2 problems with this answer.
   How did you find the problems?

8. a) Find the least 4-digit number you can subtract from 8297 without regrouping.
   b) Find the greatest 4-digit number you can subtract from 8297 without regrouping.
   c) Which number was easier to find?
      Use words and numbers to explain.

Describe 2 mental math strategies you could use to subtract 4775 − 2986.
Use words and numbers to explain.

Numbers Every Day

Number Strategies

Order the numbers in each set from least to greatest.

- 25 317, 23 715, 25 731
- 60 324, 62 043, 60 243
- 38 960, 38 906, 38 690

LESSON 6

Subtracting with 4-Digit Numbers

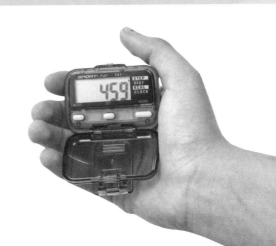

A pedometer records
the number of steps you take.

Explore

Emma wore a pedometer for 2 hours.
She recorded the number of steps each hour.
The first hour, she took 1347 steps.
The second hour, she took 984 steps.

In which hour did she take more steps?
How many more?

Show and Share

Share your work with another pair of students.
How did you subtract the numbers?

Connect

Drake and Carly went for a walk.
Each of them wore a pedometer.
After 90 minutes, Drake had taken 1276 steps
and Carly had taken 1593 steps.
How many more steps did Carly take than Drake?

Subtract: 1593 − 1276

LESSON FOCUS | Use different strategies to subtract 4-digit numbers.

➤ Use Base Ten Blocks on a place-value mat to subtract.

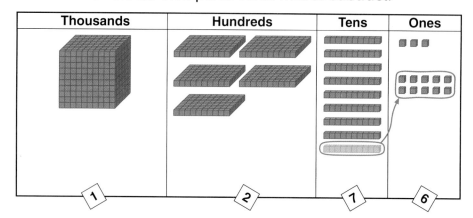

You cannot take 6 ones from 3 ones.

Trade 1 ten for 10 ones.

Take away:
- 6 ones
- 7 tens
- 2 hundreds
- 1 thousand

$1593 - 1276 = 317$

➤ Use place value to subtract.

You cannot take 6 ones from 3 ones.
Regroup 1 ten as 10 ones.

Subtract the ones.
Subtract the tens.
Subtract the hundreds.
Subtract the thousands.

$$\begin{array}{r} 15\overset{8}{\cancel{9}}\overset{13}{\cancel{3}} \\ -1276 \\ \hline \end{array}$$

$$\begin{array}{r} 15\overset{8}{\cancel{9}}\overset{13}{\cancel{3}} \\ -1276 \\ \hline 317 \end{array}$$

Carly took 317 more steps than Drake.

I add to check if the answer is correct. The sum of 1276 + 317 is 1593.

I estimate to check if the answer is reasonable. 1593 − 1276 is close to 1600 − 1300 = 300. My answer is close to 300, so it is reasonable.

Practice

Use Base Ten Blocks when they help.

1. Subtract. Check your answer.
   a) 7774
      − 1796
   b) 8350
      − 2673
   c) 6432
      − 2798
   d) 9808
      − 1750

2. Estimate first. Subtract the numbers for which the difference will be greater than 2000.
   a) 7985
      − 4512
   b) 9824
      − 7912
   c) 5632
      − 3554
   d) 7162
      − 4359

3. Mount Alberta, Alberta, is 3620 m above sea level.
   Mount Odin, Nunavut, is 2147 m above sea level.
   How much higher is Mount Alberta than Mount Odin?

4. There were 8275 tickets available for a school play.
   Two thousand eight hundred sixty-three tickets were sold.
   How many tickets were left?

5. Copy and complete the subtraction frame.
   Use the digits 1 to 9.
   Use each digit only once or not at all.
   Arrange the digits to make:
   a) the greatest difference
   b) the least difference
   How did you decide where to place the digits?

6. Suppose you subtract a 4-digit number from a different 4-digit number.
   Which numbers give the greatest difference?
   Which numbers give the least difference?
   How do you know?

Reflect

Choose one of the subtraction questions from question 1.
Explain how you found the difference.

Numbers Every Day

Calculator Skills

The product of 3 numbers is 1755.
One number is 13.
What might the other 2 numbers be?

LESSON 7: Multiplication and Division Facts to 144

Most multiplication facts have:
- one related multiplication fact
- two related division facts

$9 \times 8 = 72$
$8 \times 9 = 72$
$72 \div 9 = 8$
$72 \div 8 = 9$

Factors are numbers you multiply to get a product. 9 and 8 are factors of 72.

Some multiplication facts have:
- one related division fact

$8 \times 8 = 64$
$64 \div 8 = 8$

How do you know how many related facts a multiplication fact has?

Explore

Your teacher will give you a large copy of this multiplication chart.

Use patterns to complete the chart.
Use the completed chart to write:
- the multiplication facts with 11 as a factor
- the multiplication facts with 12 as a factor

For each of these multiplication facts, write all the related facts.

Show and Share

Share your work with another pair of students.
What patterns did you use to complete the chart?
Use the number facts to write the multiples of 12.

×	1	2	3	4	5	6	7	8	9	10	11	12
1	1	2	3	4	5	6	7	8				
2	2	4	6	8	10	12	14	16				
3	3	6	9	12	15	18	21	24				
4	4	8	12	16	20	24	28	32				
5	5	10	15	20	25	30	35	40				
6	6	12	18	24	30	36	42	48				
7	7	14	21	28	35	42	49	56				
8	8	16	24	32	40	48	56	64				
9												
10												
11												
12												

46 LESSON FOCUS | Use patterns to multiply and to divide.

Connect

Here are some strategies to help you multiply.

➤ Use known facts to multiply.
To find 12 × 8:

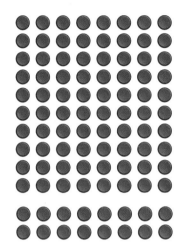

$10 \times 8 = 80$

$2 \times 8 = 16$

$80 + 16 = 96$
$12 \times 8 = 96$

➤ Use patterns to remember multiplication facts with 11.
$1 \times 11 = 11$
$2 \times 11 = 22$
$3 \times 11 = 33$
$4 \times 11 = 44$ ⟵ The digits of the number multiplied by 11 are the first and last digits of the product.
⋮
$9 \times 11 = 99$
$10 \times 11 = 110$
$11 \times 11 = 121$
$12 \times 11 = \mathbf{132}$ ⟵ The middle digit of the product is the sum of the first and last digits of the product: $1 + 2 = 3$

The products 11, 22, 33, ... 121, 132 are multiples of 11.

Here is a strategy for division.

➤ Use related multiplication facts to find the quotient.
To find $77 \div 11$:

Think: 11 times which number is 77?
You know $11 \times 7 = 77$.
So, $77 \div 11 = 7$

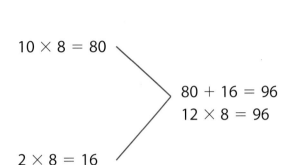

The **divisor** is 11.

$77 \div 11 = 7$

The **dividend** is 77.   The **quotient** is 7.

Practice

1. Multiply.
   a) 8 × 9
   b) 3 × 12
   c) 11 × 7
   d) 12 × 6
   e) 10 × 12
   f) 8 × 11
   g) 7 × 12
   h) 12 × 4

2. Find each quotient.
   a) 72 ÷ 6
   b) 110 ÷ 11
   c) 121 ÷ 11
   d) 33 ÷ 3
   e) 108 ÷ 12
   f) 12 ÷ 12
   g) 24 ÷ 12
   h) 55 ÷ 11

3. Write four related facts for each set of numbers.
   a) 11, 12, 132
   b) 6, 11, 66
   c) 3, 12, 36
   d) 12, 8, 96

4. a) Start at 7. Write the multiples of 7 to 84.
   b) Start at 9. Write the multiples of 9 to 108.
   c) Which multiple of 9 is also a multiple of 7? Explain.

5. Write the products for 1 × 12 to 10 × 12.
   Describe all the patterns you see in the products.
   Do the patterns continue for 11 × 12 and 12 × 12? Explain.

6. How can you use the product of 10 × 6 to help you find the product of 11 × 6?

7. Barbara planted 12 rows of corn.
   Each row had the same number of seeds.
   Barbara planted 84 seeds.
   How many did she plant in each row?

8. Kayla finds the multiplication facts for 12 by doubling the multiplication facts for 6.
   Does Kayla's strategy work?
   Use words, numbers, or pictures to explain.

Numbers Every Day

Number Strategies

Which questions have the answer 250?
How do you know?

- 775 ÷ 3
- 135 + 115
- 3875 − 3625
- 25 × 11

Reflect

Which facts do you find the hardest to remember?
Which strategies do you use to help you?
Use words, numbers, and pictures to explain.

Banquet Coordinator

World of Work

Many large restaurants, halls, and resorts cater weddings and other parties. The banquet coordinator makes sure there is the right amount of food, drink, plates, cutlery, and serving dishes for the event. If there is too much food, the restaurant can lose money. If there is too little food, the guests may complain, and the restaurant may lose future business.

The banquet coordinator works on a per person basis for most calculations. For other items, he relies on experience. For example, he knows the quantity of mineral water to order. He has extra sets of cutlery on hand in case any gets dropped. He knows how many guests 1 urn of coffee will serve.

Many establishments use spreadsheets to do the calculations. The coordinator enters the number of people attending and the spreadsheet calculates the amounts needed.

Multiplication Tic-Tac-Toe

You will need 20 each of two colours of counters and 2 paper clips. Your teacher will give you a copy of the game board and the factor list.

The object of the game is to be the first player to place 3 counters in a row. The row can be horizontal, vertical, or diagonal.

➤ Each player chooses a different colour.
➤ The first player chooses any two factors in the factor list.
She marks the factors with paper clips.
➤ Player 1 multiplies the factors.
She finds the product on the game board and covers it with a coloured marker.
If the product appears more than once on the game board, she chooses which one to cover.
➤ Player 2 may move only one of the paper clips on the factor list.
He finds the product of the factors.
He finds the product on the game board and covers it with a marker.
➤ Players continue to take turns.
Each player may move only one paper clip per turn.
➤ The first player to place 3 counters in a row wins.

Share your strategies for playing the game.
Talk about how you found products that you did not know automatically.

Variation:
Play 4-in-a-Row.

Multiplying with Multiples of 10

Every multiple of 10 has 10 as a factor.
These are multiples of 10:

    100   1000   30   300   3000

What are some other multiples of 10?

 You will need a calculator.

➤ Find each product.
Record the products in a place-value chart.

11 × 1	9 × 9	12 × 8
11 × 10	9 × 90	12 × 80
11 × 100	9 × 900	12 × 800
11 × 1000	9 × 9000	12 × 8000

Ten Thousands	Thousands	Hundreds	Tens	Ones

➤ Find each product.
Record the products in a place-value chart.

20 × 9	70 × 7	50 × 6
20 × 90	70 × 70	50 × 60
20 × 900	70 × 700	50 × 600

Show and Share

Share your work with another pair of students.
Describe any patterns you see.
How can you tell how many digits each product will have?
How can you tell which digits in a product will be 0?

LESSON FOCUS | Use patterns to multiply with multiples of 10.

Connect

➤ Use place value to multiply by 10, 100, and 1000.
Find each product. Record each product in a place-value chart.
25 × 10   25 × 100   25 × 1000

25 × 1 ten = 25 tens
25 × **10** = 25**0**

> When you multiply a whole number by 10, the digits move 1 place to the left.

25 × 1 hundred = 25 hundreds
25 × **100** = 25**00**

> When you multiply a whole number by 100, the digits move 2 places to the left.

25 × 1 thousand = 25 thousands
25 × **1000** = 25 **000**

> When you multiply a whole number by 1000, the digits move 3 places to the left.

Product	Ten Thousands	Thousands	Hundreds	Tens	Ones
250			2	5	0
2500		2	5	0	0
25 000	2	5	0	0	0

➤ Use basic facts and place-value patterns
to multiply by multiples of 10, 100, and 1000.
Find each product.
3 × 6**00**   3 × 6**000**

You know 3 × 6 = 18.

3 × 6 hundreds = 18 hundreds
3 × **6**00 = **18**00

> When you multiply a whole number by a multiple of 100, the digits in the product of the related basic fact move 2 places to the left.

3 × 6 thousands = 18 thousands
3 × **6**000 = **18** 000

> When you multiply a whole number by a multiple of 1000, the digits in the product of the related basic fact move 3 places to the left.

➤ Use what you know about
multiplying by multiples of 10, 100, and 1000
to multiply two multiples of 10, 100, and 1000.
Find each product.
20 × 30   500 × 40

2 tens × 30 = 60 tens   5 hundreds × 40 = 200 hundreds
20 × **30** = **60**0   **500** × **40** = **20** **000**

Practice

1. Multiply.
   a) 47 × 10
      47 × 100
      47 × 1000
   b) 32 × 10
      32 × 100
      32 × 1000
   c) 20 × 10
      20 × 100
      20 × 1000
   d) 50 × 10
      50 × 100
      50 × 1000

2. Use a basic fact and place-value patterns to find each product.
   a) 7 × 8
      7 × 80
      7 × 800
      7 × 8000
   b) 11 × 6
      11 × 60
      11 × 600
      11 × 6000
   c) 12 × 9
      12 × 90
      12 × 900
      12 × 9000

3. Multiply.
   a) 20 × 40
   b) 30 × 10
   c) 40 × 70
   d) 60 × 90
   e) 80 × 50
   f) 70 × 80
   g) 50 × 60
   h) 90 × 30

4. Mike works in a bank. He receives these deposits.
   How much money is in each deposit?
   a) twelve $10 bills
   b) sixty $20 bills
   c) thirty $50 bills
   d) fifteen $100 bills
   e) twenty $20 bills and ten $50 bills

5. A ruby-throated hummingbird flaps its wings about 60 times each second.
   How many times would it flap its wings in one minute? In one hour? Show your work.

6. How many seconds are there in 1 hour?

7. Write a story problem that can be solved by multiplying by a multiple of 10 000.
   Solve your problem.
   How did you solve the problem?

Reflect

Describe any patterns in the products when you multiply with multiples of 10. Use words and numbers to explain.

Numbers Every Day

Mental Math

Subtract.
Which strategies did you use?

579 − 268
1140 − 1019
483 − 268
2424 − 1212

ASSESSMENT FOCUS | Question 5

LESSON 9

Using Mental Math to Multiply

Explore

How many different ways can you find the product 14 × 26?
Record each way.

Use any materials that help.

Show and Share

Share your work with another pair of students.
Compare the strategies you used
to find the product.

Connect

➤ Multiply: 15 × 7

Think of an array for 15 × 7.

The product 15 × 7 is equal to
the sum of the products
10 × 7 and 5 × 7.

$$15 \times 7 = (10 \times 7) + (5 \times 7)$$
$$= 70 + 35$$
$$= 105$$

So, 15 × 7 = 105

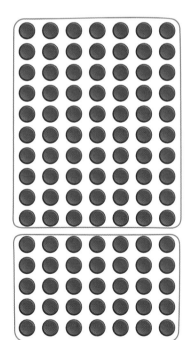

10 × 7

5 × 7

54   **LESSON FOCUS** | Use different strategies to mentally multiply 2 numbers.

➤ Multiply: 16 × 25
Use the strategy of halving and doubling.

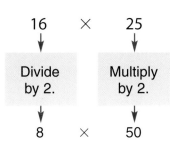

To halve is to divide by 2.
To double is to multiply by 2.

8 × 50 = 400
So, 16 × 25 = 400

➤ Multiply: 28 × 25
Rewrite 28 as 7 × 4.

28 × 25 = 7 × 4 × 25
        = 7 × 100
        = 700

So, 28 × 25 = 700

Remember:
2 × 5 = 10
4 × 25 = 100

➤ Jane has one hundred ninety-eight $5 bills.
How many dollars does she have?

Multiply: 198 × 5

Use friendly numbers.
200 is close to 198.

Multiply: 200 × 5 = 1000

200 is 2 more $5 bills than Jane has,
so subtract 2 × 5 = 10.

1000 − 10 = 990
So, 198 × 5 = 990
Jane has $990.

Practice

1. Multiply. Use mental math.
   a) 6 × 199
   b) 7 × 302
   c) 3 × 498
   d) 5 × 310
   e) 36 × 25
   f) 11 × 34
   g) 24 × 21
   h) 12 × 51

2. **Who Has the Greater Product?**

    You will need a set of shuffled digit cards from 0 to 9.
    The goal is to arrange 4 digits to make
    a multiplication problem with the greatest product.
    Each player copies and completes the multiplication grid.
    Take turns drawing one card.
    As each card is selected, each player writes that digit
    in any box on her or his grid.
    Continue until all the boxes are filled.
    Multiply.
    The player with the greater product scores a point.
    The first player to score 5 points wins.

3. Describe the strategies you used to play the game
   Who Has the Greater Product?

4. Use mental math.
   Find the product of 16 × 99 two different ways.
   Describe the strategies you used.

5. A theatre has 16 rows of seats.
   Each row has 24 seats.
   How many seats are there in the theatre?

6. Copy the multiplication frame.
   Arrange the digits 2, 3, 4, and 5 to make the greatest product.
   Use each digit only once.
   How did you decide how to arrange the digits?

7. Write a multiplication problem that can be solved
   using mental math. Solve the problem.
   Which strategy did you use? Why?

Reflect

Peter says the strategy of halving and doubling
works best when you are multiplying
even numbers.
Use words and numbers to show why
he is correct.

Number Strategies

Find each product.
Describe any patterns in the
factors and in the products.
   11 × 9
   11 × 10
   11 × 11
   11 × 12

LESSON 10

Solving Problems by Estimating

Explore

You will need counters.

Use a counter to model a quarter.
About how many quarters would cover one desk?
What would the value of the quarters be?

Show and Share

Share your results with another group.
You did not have enough counters to cover the desk.
What strategies did you use to estimate?

Connect

Here are some strategies you can use to estimate to solve problems.

➤ $873 is to be shared among 9 people.
About how much will each person get?
Estimate: 873 ÷ 9
Look for **compatible numbers**.

873 is close to 900.

9 hundreds ÷ 9 = 1 hundred
= 100

Each person will get about $100.

Compatible numbers are pairs of numbers you can divide mentally.

LESSON FOCUS | Use different strategies to estimate to solve problems.

➤ Estimate: 658 ÷ 5
  - Use front-end estimation.
    658 is about 600.
    We know that 60 ÷ 5 = 12.
    So, 600 ÷ 5 = 120

> For front-end estimation, you consider only the first digit of the number. 658 is about 600.

  - Use rounding and compatible numbers.
    658 is close to 660.
    We know that 66 ÷ 6 = 11.
    So, 660 ÷ 6 = 110
    And, 658 ÷ 5 will be a bit more than 110.

Here is a mental math strategy for division.

➤ Divide: 336 ÷ 4
  Break 336 into 2 numbers you can divide easily by 4.

  336 = 320 + 16

  320 ÷ 4 = 32 tens ÷ 4   16 ÷ 4 = 4
          = 8 tens
          = 80

  336 ÷ 4 = 80 + 4
          = 84

Numbers Every Day

Number Strategies

Find the next 3 terms in each pattern.
Write each pattern rule.

3, 6, 12, 24, …

384, 192, 96, 48, …

Practice

1. Which compatible numbers would you use to estimate each quotient?
   Why did you choose those numbers?
   a) 238 ÷ 8   b) 193 ÷ 2   c) 742 ÷ 7   d) 384 ÷ 4

2. Estimate each quotient. Which strategies did you use?
   a) 486 ÷ 5   b) 768 ÷ 7   c) 476 ÷ 8   d) 927 ÷ 9

3. Nine hundred seventy-five candies are to be shared among 9 students. About how many candies will each student get?

4. a) Nine hundred thirty bottles are placed in cartons of 6. About how many cartons are there?
   b) Eight hundred twenty-eight pencils are packaged in boxes of 8. About how many boxes are there?

5. In the photographs section of the yearbook, there are 8 student photos per page. About how many pages are needed for 654 photos?

6. Kris has 862 game tokens. He plans to share them among 9 people. About how many tokens will each person get? How did you find out?

7. How could you estimate to find out how many pennies would fill a 10-L pail? Explain what you would do. Use counters to represent pennies. About how many pennies fill a 10-L pail?

8. Twenty-two students organized a charity walk around the school. The distance is 648 m. Each student walked around once. The parent council agreed to pay $8 for each kilometre walked.
   a) Approximately how far did the students walk?
   b) About how much money did they collect?

9. One toonie is 2.80 cm wide.
   a) Fifty toonies are laid in a row. About how long is the row?
   b) What is the approximate value of 1 km of toonies placed in a row?

10. Geri bought 15 boxes of pencils. Each box contained 248 pencils.
    a) About how many pencils did she buy?
    b) Geri put the pencils in packets of 8 pencils. About how many packets were there?

11. How could you estimate how many 1-cm cubes fill a shoe box?

12. How could you estimate how many people could stand in the classroom?

Reflect

When you estimate to solve a problem, what do you do to find compatible numbers? Use an example to explain.

ASSESSMENT FOCUS | Question 7

L E S S O N 11

Multiplying Whole Numbers

Explore

You will need Base Ten Blocks or grid paper.

➤ Use Base Ten Blocks or grid paper to find the product 14 × 23. Record your work.

Show and Share

Share your results with another pair of students. Compare your diagrams.

Connect

Multiply: 21 × 13

➤ Use Base Ten Blocks. Make a rectangle that is 21 units by 13 units.

There are:
- 2 hundreds or 200
- 7 tens or 70
- 3 ones or 3

200 + 70 + 3 = 273

60 LESSON FOCUS | Use different strategies to multiply two numbers.

➤ You can also use grid paper.
Make a rectangle that is 21 units long by 13 units wide.

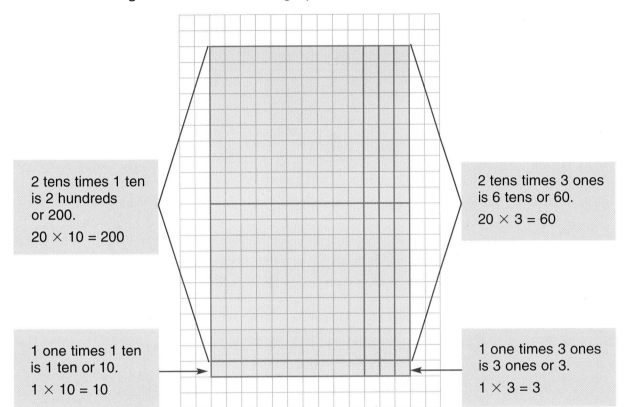

2 tens times 1 ten is 2 hundreds or 200.
20 × 10 = 200

2 tens times 3 ones is 6 tens or 60.
20 × 3 = 60

1 one times 1 ten is 1 ten or 10.
1 × 10 = 10

1 one times 3 ones is 3 ones or 3.
1 × 3 = 3

200 + 10 + 60 + 3 = 273

➤ Here is another way to explain the numbers on the grid.
Break a number apart to multiply:

```
        13
      × 21
        13
     + 260
       273
```

Multiply: 13 × 1
Multiply: 13 × 20
Add.

21 × 13 = 273

Estimate to check.

Round to friendly numbers.
13 × 21 is about 15 × 20 = 15 × 2 × 10
                       = 30 × 10
                       = 300
The estimate, 300, is close to the answer, 273.

I round 21 down to 20. So, I round 13 up to 15.

Unit 2 Lesson 11

➤ When the numbers are greater, we cannot use Base Ten Blocks
or grid paper to multiply:
Multiply: 236 × 24

```
        236
      ×  24
        944
     + 4720
       5664
```

Multiply: 236 × 4
Multiply: 236 × 20
Add.

236 × 24 = 5664
Estimate to check.

Round to friendly numbers.
236 × 24 is about 200 × 25 = 2 × 100 × 25
= 2 × 25 × 100
= 50 × 100
= 5000

The estimate, 5000, is close to the answer, 5664.

Numbers Every Day

Calculator Skills

Suppose the 8 key on your calculator is broken. How can you find each result?

870 × 8

7.8 + 8.7

Practice

1. Multiply. Use a different method to check. What do you notice about the products in each pair?

    a) 34 × 26   26 × 34
    b) 45 × 23   23 × 45

2. Multiply.
    a) 35 × 52
    b) 65 × 30
    c) 48 × 25
    d) 41 × 74
    e) 92 × 43
    f) 14 × 75
    g) 20 × 54
    h) 25 × 16

3. Find each product.
    a) 46 × 64
    b) 23 × 50
    c) 61 × 11
    d) 17 × 33
    e) 29 × 41
    f) 68 × 12
    g) 80 × 16
    h) 16 × 77
    i) 85 × 58
    j) 60 × 74
    k) 13 × 18
    l) 21 × 22

4. Can you use mental math to find any of the products in question 3? Explain.

5. Multiply. Estimate to check.
    a) 324 × 19
    b) 237 × 28
    c) 37 × 187

6. Find the product 25 × 25.
   How can you use the product 25 × 25 to help you find each product?
   a) 25 × 26   b) 24 × 25   c) 50 × 25

7. Jordan tiled a wall. His wall has 27 rows each with 27 tiles.
   Sharma tiled a different wall. Her wall has 26 rows of 29 tiles.
   Whose wall has more tiles? How many more?

8. Which multiplication facts can you use to find 45 × 23?
   How do you know?
   Show your work.

9. A working elephant eats about 225 kg of food a day.
   It drinks about 150 L of water each day.
   a) Estimate the amount of food and drink it will consume in 3 weeks.
   b) Calculate these amounts.

10. Anjotie rents out 24 kayaks for $14 per hour.
    All the kayaks are rented for 8 hours.
    How much will Anjotie earn?

11. Erica earns $9 per hour. She works 32 hours per week.
    Estimate, then calculate, how much Erica earns in 2 weeks.

12. Suppose you wanted to arrange 4 different digits to make the greatest product.
    Which arrangement would you use? Why?
    a) ☐☐☐
       × ☐
    b)   ☐☐
       ×☐☐

13. Estimate, then calculate, the cost of 36 computers at $847 each.

At Home

Reflect

How are the 3 different strategies for multiplying two numbers the same? How are they different? Use words, numbers, and pictures to explain.

Measure the dimensions of a rectangular table top to the nearest centimetre. Find the area of the table top.

ASSESSMENT FOCUS | Question 8

LESSON

Dividing Whole Numbers

Explore

A tire factory makes
824 tires a day.
A new car needs
a set of 4 tires.
How many sets of tires
are made each day?

Show and Share

Share your answer with another
pair of students.
How did you divide to find
the number of sets?

Connect

Some vehicles have 5 tires in a set.
How many sets of 5 tires can be made with 728 tires?

Divide: $5\overline{)728}$

➤ Think multiplication.
   Estimate the number of sets,
   then improve the estimate.

   728 is between 500 and 1000.
   $500 \div 5 = 100$
   $1000 \div 5 = 200$
   So, the quotient is between 100 and 200.

64 LESSON FOCUS | Use different strategies to divide a 3-digit number by a 1-digit number.

Make 100 sets of 5.	Make 40 sets of 5.	Make 5 sets of 5.
$100 \times 5 = 500$	$40 \times 5 = 200$	$5 \times 5 = 25$
Subtract: $728 - 500$	Subtract: $228 - 200$	Subtract: $28 - 25$
		Add the sets.

```
   5)728              5)728              5)728
   -500  100          -500  100          -500  100
   ----               ----               ----
    228                228                228
                      -200   40          -200   40
                      ----               ----
                        28                 28
                                          -25   + 5
                                          ---   ----
                                            3    145
```

So, $728 \div 5 = 145$ R3

➤ Use place value.

Show 7 hundreds in 5 equal groups. There are 5 groups each with 1 hundred. 2 hundreds are left over.	2 hundreds 2 tens are 22 tens. Show 22 tens in 5 equal groups. There are 5 groups each with 4 tens. 2 tens are left over.	2 tens 8 ones is 28 ones. Show 28 ones in 5 equal groups. There are 5 groups each with 5 ones. 3 ones are left over.

```
   h t o              h t o              h t o
   1                  1 4                1 4 5
  5)7 2 8            5)7 2 8            5)7 2 8
   -5                 -5↓                -5↓
   ---                ---                ---
    2                  2 2                2 2
                      -2 0               -2 0↓
                      ---                ---
                         2                 2 8
                                          -2 5
                                          ---
                                             3
```

So, $728 \div 5 = 145$ R3

One hundred forty-five sets of tires can be made.
There will be 3 tires left over.

To check, multiply 145 by 5, then add 3.
$145 \times 5 = 725$
$725 + 3 = 728$ ⟵ Since this is the dividend, the answer is correct.

➤ Use short division.

$5\overline{)7^22\,8}$
  1

7 hundreds ÷ 5 = 1 hundred, with 2 hundreds left over

$5\overline{)7^22^28}$
 1 4

22 tens ÷ 5 = 4 tens, with 2 tens left over

$5\overline{)7^22^28}$
 1 4 5 R3

28 ones ÷ 5 = 5 ones, with 3 ones left over

Practice

1. Find each quotient. Estimate first.
   a) 9)540
   b) 3)720
   c) 5)255
   d) 8)168
   e) 4)268
   f) 7)112
   g) 6)704
   h) 2)173
   i) 9)398
   j) 4)600
   k) 3)299
   l) 3)212

2. Divide. Check by multiplying.
   a) 925 ÷ 6
   b) 537 ÷ 9
   c) 588 ÷ 7
   d) 831 ÷ 4
   e) 108 ÷ 4
   f) 311 ÷ 6
   g) 284 ÷ 5
   h) 606 ÷ 9
   i) 667 ÷ 7
   j) 424 ÷ 8
   k) 903 ÷ 8
   l) 418 ÷ 6

3. Most minivans use 3 windshield wiper blades.
   How many sets of 3 blades can be made from 342 blades?

4. Gabi has 629 pennies.
   She wants to share them equally among 7 people.
   a) Estimate how many pennies each person gets.
   b) Calculate how many pennies each person gets.
      How did you find out?

5. Zoomin' Inc. makes skateboards.
   In 5 days, 980 skateboards were made.
   The same number of skateboards was made each day.
   How many skateboards were made each day?
   How can you check?

6. Write a division problem that can be solved by dividing a 3-digit number by a 1-digit number.
Trade problems with a classmate.
Solve your classmate's problem.

7. **Target No Remainder!**
You will need:
   - a spinner with 6 equal sectors, labelled 4 to 9
   - 3 number cubes, each labelled 1 to 6

   Take turns.
   On your turn, roll all 3 number cubes and spin the spinner.
   Arrange the numbers on the number cubes to make a 3-digit number.
   Divide the 3-digit number by the number on the spinner.
   Record the remainder. This is your score for this turn.
   At the end of the game, total your score.
   The winner is the player with the lesser total.

8. Use the digits 8, 6, and 1 once each.
Arrange the digits to make a 3-digit number.
How many different 3-digit numbers can you make that are divisible by 7 with no remainder?
How do you know you have found all of them?

9. Without dividing, how can you tell how many digits the quotient of 844 ÷ 9 will have?
Use words and numbers to explain.

Reflect

Look at your answers for question 2.
Which quotients had 3 digits? Which had 2 digits?
How can you tell how many digits the quotient will have before you divide?

Numbers Every Day

Mental Math

Find each sum.

8 + 15 + 12
41 + 17 + 3 + 9
38 + 12 + 25

Which strategies did you use?

LESSON 13

Solving Problems

You have used addition, subtraction, multiplication, and division to solve problems with whole numbers.

In this lesson, you will solve problems with more than one step.

Explore

Rhianna mows lawns and shovels driveways.
Last year, she earned $1252.
She mowed 93 lawns for $8 each.
How much money did she earn
from shovelling driveways?

Show and Share

Share your work with another pair of students.
Compare your answers and
the strategies you used to find them.
What did you need to calculate before
you could find how much Rhianna earned
from shovelling driveways? Explain.

Connect

➤ Rob spent $1478 on stamps and coins.
He bought 14 stamps for $37 each.
How much did Rob spend on coins?

To find the amount Rob spent on coins,
we need to find out how much Rob spent
on stamps.

Multiply: 14 × 37

$$\begin{array}{r} 37 \\ \times\ 14 \\ \hline 148 \\ +\ 370 \\ \hline 518 \end{array}$$

Rob spent $518 on stamps.

Find how much Rob spent on coins.
Subtract the amount he spent on stamps
from the total amount he spent.
Subtract: 1478 − 518

1478 − 518 = 960

Rob spent $960 on coins.

➤ Mackenzie uses 16 m of fabric to make 4 outfits
from one pattern.
How much fabric would she need to make 9 outfits
from the same pattern?

To find the amount of fabric she needs for 9 outfits,
we first need to know how much fabric she needs for 1 outfit.
Divide: 16 ÷ 4 = 4

Mackenzie needs 4 m of fabric to make 1 outfit.
Multiply the amount of fabric needed for 1 outfit
by the number of outfits, 9.
4 × 9 = 36

Mackenzie needs 36 m of fabric to make 9 outfits from the pattern.

Practice

1. Campbell bought 148 hardcover books. Each book cost $35.
   a) How much did Campbell spend on books?
   b) Write a story problem that uses your answer to part a.
      Trade problems with a classmate.
      Solve your classmate's problem.
   c) Compare your problem to your classmate's problem.

2. For each problem, describe what you need to find before you can solve the problem.
   a) At Sam's Office Supply, a package of 3 colour inkjet cartridges costs $216.
      At Ink World, the same brand of cartridge costs $79 each.
      How much more does a colour cartridge cost at Ink World?
   b) Karen booked the computer for 2 hours.
      She spent 75 minutes typing a report and 32 minutes checking her work.
      How much computer time does Karen have left?

3. The Lakeland District choir stood in rows of 12 for a performance.
   The people in 2 rows carried red streamers.
   The people in 4 rows carried yellow streamers.
   The people in 3 rows carried purple streamers.
   How many people are in the choir?

4. Connor runs 150 m every minute.
   A cheetah runs 29 m every second.
   a) How much farther than Connor will the cheetah run in 1 minute?
   b) Explain how you solved the problem.

5. Kamil played a game 3 times.
   His first score was 1063 points.
   His second score was 129 points lower.
   His third score was 251 points higher than his second score.
   How many points did Kamil score in his third game?

Reflect

Describe how to solve a problem with more than one step.
Use examples to explain.

Numbers Every Day

Number Strategies

Find each product.

25 × 32
47 × 50
22 × 88
80 × 60
14 × 19

Less Is More

You will need a decahedron numbered 0 to 9.

The object of the game is to make a division statement with:
- the least quotient, *and*
- the least remainder

➤ Each player makes a large copy of this division frame.

☐ ☐ ☐ ÷ ☐

➤ Players take turns to roll the decahedron.
➤ On your turn, record the number that turns up in any square of the division frame.
Once a number is written down, you may not move it.
➤ Continue until each player has filled her or his division frame.
➤ Each player finds the quotient for her or his division frame.
Check each other's work.
Each player with a correct answer scores 1 point.
The player with the least remainder scores 1 point.
The player with the least quotient scores 1 point.
The first player to score 6 points wins.

LESSON 14

Strategies Toolkit

Explore

Samrina organized a team to participate in a 325-km bike relay.
Half the team members ride 25 km.
The rest ride 40 km.
Including Samrina, how many people are on Samrina's team?

Show and Share

Describe the strategy you used to solve the problem.
How could you solve the problem a different way?

Connect

Mr. Biegala bought resource books for $28 and bookshelves for $84.
He spent $448 on 12 items.
How many of each item did Mr. Biegala buy?

Strategies
- Make a table.
- Use a model.
- Draw a diagram.
- Solve a simpler problem.
- Work backward.
- Guess and check.
- **Make an organized list.**
- Use a pattern.
- Draw a graph.

What do you know?
- Resource books cost $28 each.
- Bookshelves cost $84 each.
- The total number of books and bookshelves is 12.
- The total cost is $448.

Think of a strategy to help you solve the problem.
You could **make an organized list**.
- Find 2 numbers that add to 448.
- One number must be a multiple of 28, the other must be a multiple of 84.

LESSON FOCUS | Interpret a problem and select an appropriate strategy.

Find the cost of 1 bookshelf.
Find the amount of $448 that is left.
Try to find a number of books whose cost is equal to the amount left.
Record the amounts in an organized list.
Find the cost of 2 bookshelves.
Continue until you find the number of books and the number of bookshelves.

Number of Bookshelves	Cost ($)	Amount Left ($)	Number of Books	Cost ($)
1	84	448 − 84 = 364	11	308

Check your work.
Is the total number of books and bookshelves 12?
Is the total cost of books and bookshelves $448?

Practice

Choose one of the **Strategies**

1. Together, 2 hot tubs cost $6094.
   The difference between their prices is $2304.
   What is the price of the more expensive hot tub?

2. Colin's grandma gave him $100.
   He bought a game for $61.
   He wants to buy another game that costs $47.
   How much more money does Colin need?

Reflect

Choose one of the *Practice* questions.
Describe how you solved the problem.

Unit 2 Show What You Know

LESSON

1

1. Write each number in standard form.
   a) 800 000 + 60 000 + 400 + 30 + 7
   b) seven hundred fifty-four thousand eight

2. Write the value of the underlined digit.
   Then write the number in expanded form.
   3<u>2</u>0 075

3. Write the numbers in order from greatest to least:
   437 126   437 162   473 126

2

4. Draw rectangles on grid paper to find out if each number is prime or composite.
   a) 36   b) 37   c) 43   d) 51

3 4 5 6

5. Estimate first. Then find each sum or difference.
   a)   4729   b)   2968   c)   9127   d)   6496
      + 3178   − 387   − 6988   + 7827

6. Use mental math to add or subtract.
   a)   6647   b)   8000   c)   2999   d)   7642
      + 1004   − 3594   + 3998   − 3998

7

7. Write four related facts for each set of numbers.
   a) 8, 12, 96   b) 11, 9, 99

8. Find each product or quotient.
   a) 132 ÷ 12   b) 10 × 11   c) 108 ÷ 2   d) 7 × 11

8 9 10

9. Multiply.
   a) 8 × 7000   b) 50 × 90   c) 11 × 500   d) 60 × 60
   e) 32 × 75   f) 51 × 78   g) 661 × 33   h) 91 × 23

10. Jacob has two hundred ninety-seven $5 bills. How much money does he have?

11. Use mental math to multiply or divide.
   a) 11 × 54
   b) 3 × 498
   c) 567 ÷ 9
   d) 18 × 25

12. Sandra bought 17 CDs for $23 each. How much did she spend on CDs?

13. Divide, then check.
   a) 5)625
   b) 338 ÷ 2
   c) 4)750
   d) 382 ÷ 8

14. Bedding plants are sold in trays of 6. How many trays are needed to hold 342 plants?

15. At Marg's Market, you can buy 6 boxwood plants for $354. At Green Gardens, the same size of boxwood plant costs $53. Which store has the better price on boxwood plants? How do you know?

16. An apartment building has 32 one-bedroom apartments, 24 two-bedroom apartments, and 16 three-bedroom apartments. How many bedrooms are in the building?

UNIT 2 Learning Goals

- ✓ read and write numbers to 999 999
- ✓ estimate to solve problems
- ✓ compare and order numbers
- ✓ use place value to represent numbers
- ✓ find prime and composite numbers
- ✓ find factors and multiples
- ✓ estimate sums, differences, products, and quotients
- ✓ add, subtract, and multiply numbers mentally
- ✓ add and subtract 4-digit numbers
- ✓ multiply a 3-digit number by a 2-digit number
- ✓ divide a 3-digit number by a 1-digit number
- ✓ solve problems using whole numbers
- ✓ solve problems with more than one step

Unit 2 **75**

Unit Problem

On the Dairy Farm

Silage is made from green corn plants. The whole plant is harvested, chopped, and fermented in a storage silo.

Haylage is hay that has been cut, chopped, and stored moist.

Each day a cow eats:
- 5 kg of hay
- 9 kg of haylage
- 9 kg of corn silage
- 10 kg of dairy ration

A cow also requires minerals and salt, and 80 to 160 L of water each day.

Check List

Your work should show
- ☑ that you can choose the correct operation to solve problems
- ☑ all calculations in detail
- ☑ a challenging story problem using whole numbers
- ☑ a clear explanation of how you solved your story problem

1. Amy has 43 dairy cows on her farm.
   How many kilograms of feed will she use each day? Every 2 weeks?

2. Matthew has 72 hectares of field on his farm.
   He plans to use 4 parts to plant hay, 1 part to plant corn, and 1 part as cow pasture.
   How many hectares of field will he use for each purpose?

   1 hectare is equal to 10 000 m².

3. It takes a milking machine about 5 minutes to milk a cow.
   The Colyns can milk 14 cows at a time in their milking parlour.
   How long will it take the machine to milk all 30 cows?

4. Write a story problem about a dairy farm.
   Solve your problem.
   How did you solve the problem?

Reflect on the Unit

Describe the mental math strategies you use to add, subtract, multiply, and divide.
Give an example for each strategy.

Unit 2   **77**

UNIT 3

Geometry

Bridges

Pratt Truss

Double Warren Truss

Howe Truss

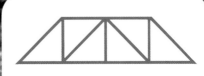
Howe Truss with counter braces

Learning Goals

- identify and name polygons
- construct and classify triangles
- build and describe figures
- identify the least information needed to draw a figure
- draw solids
- build and describe solids
- identify planes of symmetry

Key Words

polygon

quadrilateral

equilateral triangle

isosceles triangle

scalene triangle

right angle

plane

plane of symmetry

These are different types of truss bridges.

They were built during the great age of trains, about a hundred years ago.

A truss is a framework.

It is made of wooden beams or metal bars.

The bridges are light, strong, and rigid.

- What is the most common geometric figure you see in the bridges?
  How many triangles can you count in each bridge?
  How are the triangles the same?
  How are they different?
- What other geometric figures do you see?
  How are they the same? How are they different?
- Which bridge do you think would support the greatest mass? Why?

LESSON 1

Identifying and Naming Polygons

A **polygon** is a closed figure with 3 or more sides.

A **quadrilateral** is a polygon with 4 sides, 4 angles, and 4 vertices.

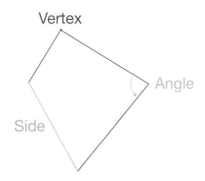

Explore

You will need a geoboard, geobands, and dot paper.

➤ Make as many different figures as you can, that have 4 sides. Record each figure on dot paper. Write how many sides, angles, and vertices it has.
What attributes does each figure have?

➤ Repeat the activity for figures with 5 sides.

➤ Continue to make figures with 6 sides, 7 sides, and so on. Draw and describe each figure you make.

Show and Share

Did any of your figures have line symmetry? How do you know?
Did any of your figures have all sides equal and all angles equal? How do you know?

Numbers Every Day

Number Strategies

Write three different number patterns that begin with 10, 20, ...
Write a rule for each pattern.

80 LESSON FOCUS | Use numbers of sides and angles to name polygons.

Here are some ways to identify and name polygons.

➤ Name polygons by the numbers of sides, angles, or vertices.
A polygon has equal numbers of sides, angles, and vertices.
So, one way to name a polygon is by the number of its sides.

| A pentagon has 5 sides. | A hexagon has 6 sides. | An octagon has 8 sides. |

➤ Name polygons by their vertices.
Label each vertex with a different capital letter.

This is triangle ABC.   This is quadrilateral MNPQ.

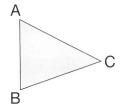 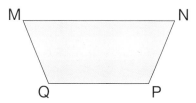

Use the letters to name the sides of the polygon.
Triangle ABC has 3 sides: AB, AC, and BC

➤ A regular polygon has all sides equal and all angles equal.
A regular polygon has lines of symmetry.

To check if all the angles are equal, trace one angle and see if it matches the other angles.

Math Link

Your World

A house is built with different figures.
A roof has triangular frames.
Windows and doors are rectangular.

A regular quadrilateral is a square.
It has 4 lines of symmetry.

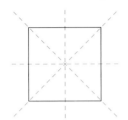

A regular pentagon has 5 lines of symmetry.

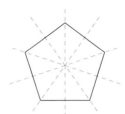

To check if a figure has a line of symmetry, trace the figure and fold it.

Practice

1. **a)** Name each polygon.

  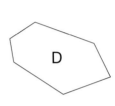 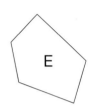

   b) Which polygons above are regular? How do you know?

2. Which polygons in question 1 have 1 line of symmetry? More than 1 line of symmetry? How do you know?

3. **a)** Name each polygon.

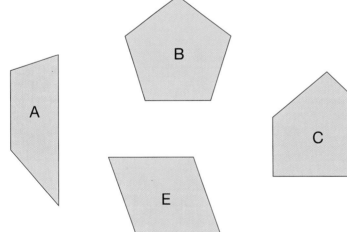
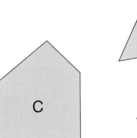

   b) Sort the polygons by number of sides.
   c) Sort the polygons by number of vertices.
   d) Compare the two sortings. What do you notice? Do you think this is always true? Explain.

4. Sailors use nautical flags to send messages.
   Flags can warn of danger
   and give weather information.
   Look at the nautical flags.
   a) List all the polygons you can find in each flag.
   b) Sketch 5 different polygons that you found.
   c) Which polygon is the most common?
      How many sides does it have?
      Explain why this might be.

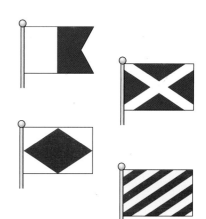

5. Use the clues to help you find the unknown attribute.

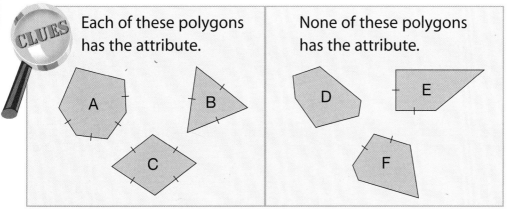

   a) What is the attribute?
   b) Which of these polygons has the attribute? Show your work.

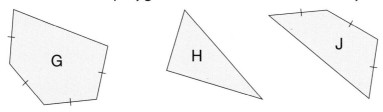

6. Use dot paper. Draw each figure.
   How many different figures can you draw each time?
   a) a triangle with exactly 2 equal sides
   b) a quadrilateral with exactly 3 equal sides
   c) a pentagon with exactly 4 equal sides
   d) a hexagon with exactly 5 equal sides

Reflect

When you see a polygon, how do you identify it?
Give at least 3 examples.

LESSON 2

Constructing Triangles

Explore

You will need a geoboard, geobands, and dot paper.
➤ Make 4 different triangles on your geoboard.
➤ Draw each triangle on dot paper.
➤ Describe each triangle as many ways as you can.

Show and Share

Share your triangles with another student.
How are your triangles alike?
How are they different?

Connect

➤ Triangles can be named by the number of equal sides.

An equilateral triangle has all sides equal.

An isosceles triangle has 2 sides equal.

A scalene triangle has no sides equal.

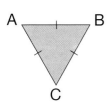
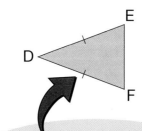
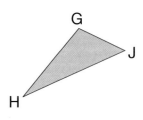

On a figure, we use hatch marks to show equal sides.

84   **LESSON FOCUS** | Classify triangles by side measures.

Practice

1. Use a geoboard, geobands, and square dot paper.
   a) Make 3 different scalene triangles.
      Record each triangle on dot paper.
      How do you know each triangle is scalene?
   b) Make 3 different isosceles triangles.
      Record each triangle on dot paper.
      How do you know each triangle is isosceles?
   c) Try to make an equilateral triangle.
      What do you notice?

2. a) Measure the sides of each triangle.

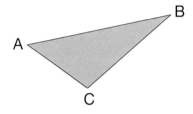 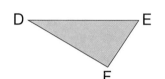 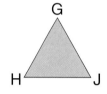

   b) Name each triangle as equilateral, isosceles, or scalene.
   c) Which triangles have lines of symmetry? How do you know?

3. Which team pennants are isosceles triangles? Explain.

   a)    b)

   c) 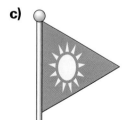   d)

Numbers Every Day

Calculator Skills

Find 3 odd numbers that have a product of 693 and a sum of 27.

Unit 3 Lesson 2 **85**

4. a) Which triangles are isosceles? How do you know?

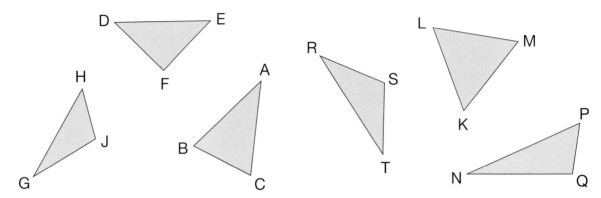

  b) For each triangle, name the sides that have the same length.
  c) Find the perimeter of each triangle.

5. You need drinking straws, scissors, and pipe cleaners.
Cut the straws into 8 pieces as shown.
Use pieces of pipe cleaner as joiners.

  a) Make each triangle.
  Trace and label your results.
  - an equilateral triangle
  - the isosceles triangle with the least perimeter
  - the scalene triangle with the greatest perimeter

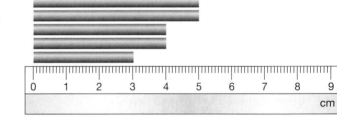

  b) Which straws could not be used together to make a triangle? Explain.

6. Use a geoboard, geobands, and dot paper.
  a) Make an isosceles triangle. Record the triangle on dot paper.
  b) Use the triangle from part a.
  Change the triangle so it is scalene.
  Describe the changes you made.

Reflect

How can you use side measures to name a triangle?
Use words and pictures to explain.

LESSON 3

Combining Figures

The size of an angle can be described by comparing it with a right angle.

This is a right angle.   This angle is less than a right angle.   This angle is greater than a right angle.

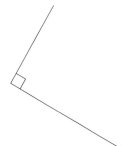
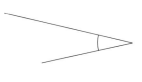

We can use these measures to describe the angles in any polygon.

Explore

You will need a tangram and dot paper.
Use 2 tans each time.
➤ Make:
  • a square
  • a rectangle
  • a parallelogram
  • a trapezoid
  • a triangle
  • your choice of figure
➤ Draw each figure on dot paper.
➤ Describe the attributes of each figure. Include angle measures.

Show and Share

Share your figures with another pair of classmates.
Was there any figure you could not make with 2 tans? Explain.

LESSON FOCUS | Use geometric figures to build other figures.

Connect

Here are some figures we can make by combining 2 Pattern Blocks.

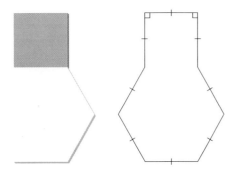

This is an octagon.
All sides are equal.
Two angles are right angles.
Six angles are greater than a right angle.

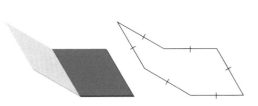

This is a hexagon.
All sides are equal.
Two angles are less than a right angle.
Four angles are greater than a right angle.

Practice

1. Use a tangram.
   Make a parallelogram with:
   a) 3 tans
   b) 4 tans
   c) 5 tans
   Draw each parallelogram on dot paper.
   Show the tans you used.

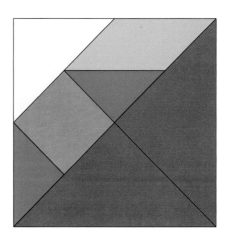

2. a) How many different polygons can you make with these 4 tans?

   b) Sketch 3 of the polygons you made.
   Show the pieces you used.
   Describe the angles in each polygon.
   Mark the equal sides.

3. Use tans to make each picture. Record your answers.
   a) Use 5 tans.   **b)** Use 7 tans.

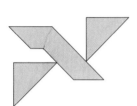 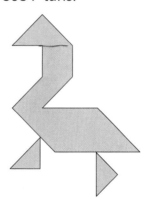

4. Work in a group. Use a tangram.
   Your teacher will give you a copy of this table.

	Triangle	Square	Rectangle	Trapezoid	Parallelogram
1 Tan					
2 Tans					
3 Tans					
4 Tans					
5 Tans					
6 Tans					
7 Tans					

Try to make each figure using the number of tans shown.
   a) Draw each figure in the table.
   Fill in as many spaces as you can.
   b) Which spaces could not be filled in? Explain.
   c) Could some spaces be filled in more than one way? Explain.

Math Link

History

Most people believe the tangram originated in ancient China. Tangrams are still very popular today. They are used to investigate important mathematical ideas. Children and adults use them to make figures and pictures.

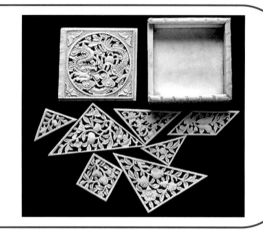

5. Use these tans:

Try to make a triangle, a square, a rectangle, a parallelogram, a trapezoid, and a pentagon.
Sketch and name each figure you make.
Is there any figure you could not make? Explain.
Show your work.

6. Use a tangram.
   a) Make a picture with some or all of the tans.
      Trace the outline of your picture.
   b) Trade drawings with a classmate.
   c) Use the tans to make your classmate's picture.

Reflect

Show how to combine polygons to make another polygon.
Use words and pictures to describe 3 different polygons made this way.

Numbers Every Day

Number Strategies

Pick any 4 consecutive numbers. Add the first and the last number. Multiply their sum by 2. Add all four numbers. What do you notice? Check to see if this works with other sets of 4 consecutive numbers.

LESSON 4

What Makes a Figure?

Congruent figures have the same size and shape.
These quadrilaterals are congruent.

Explore

You will need several straws with lengths 7 cm,
5 cm, and 4 cm, pipe cleaner pieces, and a ruler.

➤ Make four different triangles,
using 3 straws each time.
Sketch each triangle you make.
Label each side length.
How are your triangles different from
those of your partner?
How are they the same?

➤ Make a parallelogram with 4 straws.
Sketch the parallelogram.
Label each side length.
Compare your parallelogram with that of your partner.
How is your parallelogram like your partner's?
How is your parallelogram different from your partner's?

Show and Share

Look at both sets of triangles that you and your partner made.
Which triangles are congruent? How do you know?
Suppose both of you made parallelograms using 7-cm
and 4-cm straws.
Would the parallelograms be congruent? Explain.

LESSON FOCUS | Identify the least information needed to draw a figure.   **91**

Connect

➤ If we know 2 sides of a triangle, we can draw many different triangles. Here are 3 different triangles with side lengths 4 cm and 3 cm.

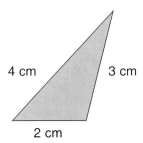 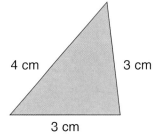 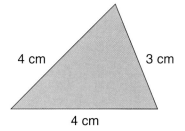

➤ If we know 3 sides of a triangle, we can draw only one triangle. Here is the only triangle we can draw with side lengths 3 cm, 4 cm, and 5 cm.

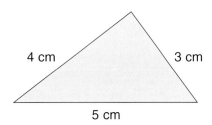

➤ If we know 2 different sides of a rectangle, we can draw only one rectangle. We know all the angles are right angles. Here is the only rectangle we can draw with side lengths 4 cm and 3 cm.

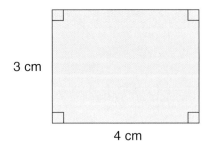

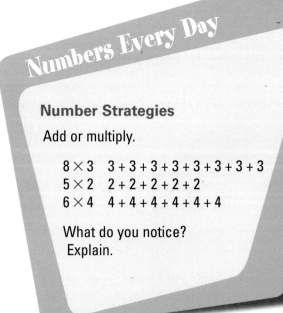

Numbers Every Day

Number Strategies

Add or multiply.

8×3 $3 + 3 + 3 + 3 + 3 + 3 + 3 + 3$
5×2 $2 + 2 + 2 + 2 + 2$
6×4 $4 + 4 + 4 + 4 + 4 + 4$

What do you notice? Explain.

➤ If we know 2 different side lengths of a parallelogram, we can draw many different parallelograms. Here are 3 different parallelograms with side lengths 4 cm and 3 cm.

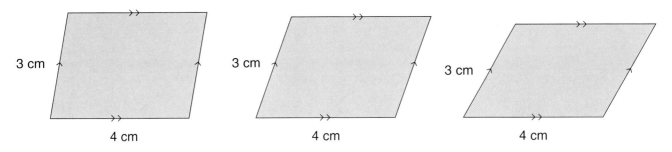

1. You know that a triangle has two sides of length 10 cm and 12 cm. What else do you need to know to make sure you and a classmate draw congruent triangles?

2. What do you need to know about a square to make sure you and a classmate draw congruent squares?

3. Use the clues to help you name each quadrilateral. How many different quadrilaterals can you name each time?
   a) I have two pairs of parallel sides.
      All my sides are equal.
      What am I?
   b) I have two pairs of parallel sides.
      I have two pairs of equal sides.
      None of my angles is a right angle.
      What am I?
   c) I have two pairs of equal sides.
      I do not have parallel sides.
      What am I?
   d) I have at least 1 pair of parallel sides.
      What am I?
   e) I am a parallelogram.
      I have 1 right angle.
      What am I?

Unit 3 Lesson 4

4. Make up your own set of clues for a quadrilateral or a triangle.
   Try to make as few clues as possible.
   Give your puzzle to a classmate to solve.

5. Use square dot paper.
   Draw and name a quadrilateral with exactly:
   a) one line of symmetry
   b) two lines of symmetry
   c) four lines of symmetry
   Try to draw more than one type of quadrilateral each time.
   Write about what you found out.

6. You can probably identify each figure below by sight.
   But, what do you need to know about each figure
   to be sure it is what you think it is?

   a)  b)  c)  d)

7. Is each statement true or false? Explain.
   a) A square is also a rhombus.
   b) A rectangle is also a square.
   c) A square is also a rectangle.
   d) A rhombus is also a parallelogram.

Reflect

Choose two figures.
What are the fewest measurements you need to know so you and a classmate draw congruent figures each time?

At Home

Road signs are polygons.
As you travel to and from school, which signs do you see?
Which polygon is used for each sign?

LESSON 5

Drawing Solids

Identify each solid. Describe its attributes.

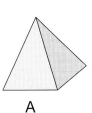
A

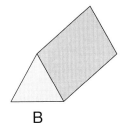
B

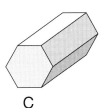
C

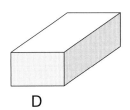
D

Explore

You will need both triangular and square grid paper and dot paper, and models of the solids above.

➤ Match each solid above with its front face below.
  Explain how you know.

E

F

G

H

➤ Choose one front face and matching solid.
  Use grid paper or dot paper.
  Sketch the solid.
➤ Trade sketches with your partner.
  Identify your partner's solid.

Show and Share

How did the dot paper or grid paper help you draw the solid?
What clues did you use to identify your partner's solid?

LESSON FOCUS | Draw a solid, given its front face.

Here are 2 ways to sketch a solid.

➤ To draw a triangular prism on triangular dot paper:

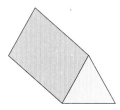

On triangular dot paper, each dot is the same distance from the 6 closest dots around it.

Step 1: Use a triangular base as the front face. Join dots to draw a triangle.

Step 2: Draw a congruent triangle that is up and to the left of the first triangle.

Step 3: Join corresponding vertices for the edges of the prism. These edges are parallel.

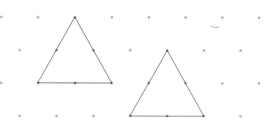

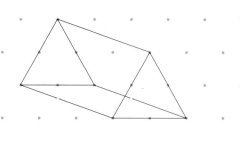

➤ To draw a square pyramid on square dot paper:

Step 1: The base is a square, but we draw a parallelogram for the base.

Step 2: Draw the diagonals of the base with broken lines.
The diagonals meet at the midpoint of the parallelogram.

Step 3: Mark a point directly above the midpoint. This new point is the top vertex of the pyramid.
Join this vertex to each vertex of the parallelogram.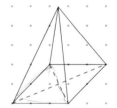

Practice

Use square dot paper or triangular dot paper.

1. Follow the steps in *Connect* to draw a triangular prism.

2. Follow the steps in *Connect* to draw a square pyramid.

3. Use the front face of each solid. Draw the solid.
    a) a rectangular prism
    b) a triangular pyramid
    c) a cube
    d) a rectangular pyramid

Unit 3 Lesson 5

4. Draw as many prisms and pyramids as possible that have a triangle as a front face.
   Write about each solid you draw.

5. Use 12 toothpicks and as many marshmallows as you need.
   The toothpicks are the edges of a solid.
   The marshmallows are the vertices of a solid.
   How many different prisms and pyramids can you make?
   Sketch and identify each skeleton you make.

6. A triangular prism with its base on the desk has this front face:

   Draw this prism.

7. Here is the front face of a rectangular prism that measures 3 cm by 6 cm by 4 cm.
   Draw this prism.

   3 cm | 6 cm

8. Name 3 objects outside the classroom that have:
   a) the shape of a prism
   b) the shape of a pyramid
   Describe each object in as much detail as possible.

Reflect

How would you explain to someone how to draw a triangular pyramid?
Write the steps. Include a drawing.

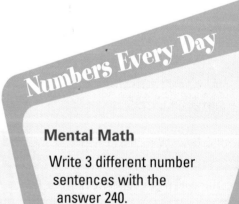

Mental Math

Write 3 different number sentences with the answer 240.

LESSON 6

Planes of Symmetry

Explore

You will need modelling clay, a ruler, and dental floss.

➤ Make several congruent cubes.
  Use the dental floss to cut the cubes.
  How many different ways can you cut
  the cubes to make two congruent parts?
  Record each way you find.
  Sketch the congruent parts.

➤ Make several congruent square pyramids.
  How many different ways can you cut
  the pyramids to make two congruent parts?
  Record each way you find.
  Sketch the congruent parts.

Show and Share

Compare your congruent parts and sketches with those of another pair of students. How many ways did you find each time? What is the shape of the face formed where the cut was made?

Connect

➤ When we cut a figure into 2 congruent parts, so one part is the mirror image of the other part, the cut line is a line of symmetry.

LESSON FOCUS | Build solids and identify their planes of symmetry.

➤ When we cut a solid into 2 congruent parts, so one part is the mirror image of the other part, the cut makes a **plane**, which is called a **plane of symmetry**.

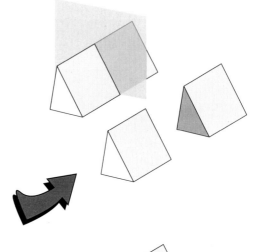

For the triangular prism in the position above, the plane of symmetry is vertical. The plane cuts the prism and makes 2 congruent triangular faces. This triangular prism has only 1 plane of symmetry.

If we cut the triangular prism horizontally, the two parts are *not* congruent. The horizontal plane is *not* a plane of symmetry.

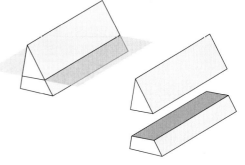

➤ A square pyramid has 4 planes of symmetry.

     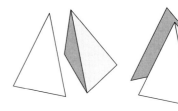

Two vertical planes cut through the midpoints of opposite sides. Each plane makes 2 congruent triangular faces.

Two vertical planes cut through the diagonals of the base. Each plane makes 2 congruent triangular faces.

Practice

1. Use modelling clay, a ruler, and dental floss. Make a rectangular prism. Find how many planes of symmetry it has. Draw the faces each plane makes where it cuts the prism.

2. Name each prism and pyramid.
   How many planes of symmetry does each solid have?
   Sketch the faces made by the plane.
   a)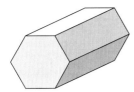
   b)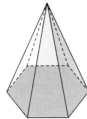

3. How many planes of symmetry does each object have?
   Sketch the faces made by each plane.
   a)
   b)

4. Use modelling clay, a ruler, and dental floss.
   Make a triangular pyramid that has equilateral triangle faces.
   a) How many planes of symmetry does this triangular pyramid have?
   b) How do you know you have found all the planes?
   c) Sketch the pyramid. Sketch each plane of symmetry and the faces it makes where it cuts.

5. A horizontal plane is drawn through a square pyramid.
   Why is this plane not a plane of symmetry?
   Include a sketch in your answer.

6. When would a horizontal plane drawn through a triangular prism be a plane of symmetry?
   Include a sketch in your answer.

Reflect

When you see a solid, how can you find how many planes of symmetry it has?
Use pictures in your explanation.

Numbers Every Day

Mental Math

Estimate each sum.

19 + 48 =
349 + 129 =
299 + 69 =

Which strategies did you use?

ASSESSMENT FOCUS   Question 4

LESSON 7: Strategies Toolkit

Explore

You will need straws, pipe cleaners, and scissors. Make skeletons of six different pyramids. How are the numbers of vertices, faces, and edges related?

A skeleton does not have faces. Use a model of the pyramid to count the faces, if you need to.

Show and Share

Compare the relationship you find, with that of another pair of classmates.

Connect

How are the numbers of vertices, faces, and edges in different prisms related?

What do you know?
- Each prism has a fixed number of faces, edges, and vertices.

Think of a strategy to help you solve the problem.
- You can **make an organized list**.
- Begin with the prism that has the fewest edges on its base.

Strategies
- Make a table.
- Use a model.
- Draw a diagram.
- Solve a simpler problem.
- Work backward.
- Guess and check.
- Make an organized list.
- Use a pattern.
- Draw a graph.

102   LESSON FOCUS | Interpret a problem and select an appropriate strategy.

Count the numbers of vertices, faces, and edges of a triangular prism, a rectangular prism, a pentagonal prism, and a hexagonal prism. Record each number in an organized list.

Continue the list.
Look for patterns.
Try adding numbers
in 2 columns, and
comparing the sum
with the number in the third column.

Prism	Vertices	Faces	Edges
triangular	6	5	9

Does the relationship in *Connect* match the relationship in *Explore*? Explain.
How can you check if the relationship is true for an octagonal pyramid and octagonal prism?

Practice

Choose one of the **Strategies**

1. Check to see if the relationship is true for these solids. What did you find out?

   a) 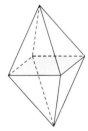　　b) 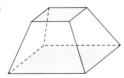　　c)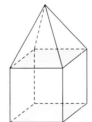

2. A solid has 7 vertices and 15 edges.
   a) How many faces does it have?
   b) What might the solid look like?

Reflect

How did you use an organized list to solve a problem?
Use an example to explain.

Unit 3 Show What You Know

LESSON

1

1. a) Name each polygon.

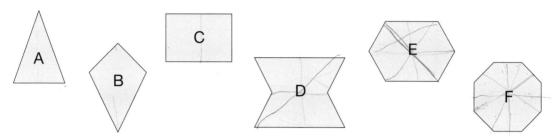

b) Which polygons above have 1 line of symmetry? More than 1 line of symmetry?

c) Which polygons above are regular? How do you know?

2. Is a rhombus a regular pentagon? How do you know?

2

3. Name each triangle as scalene, isosceles, or equilateral. Tell how you know.

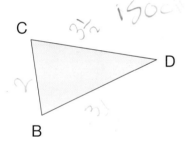 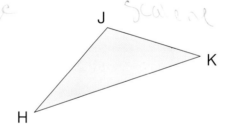 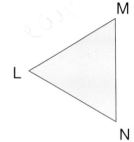

4. Draw this square on dot paper.
   Draw the line segments shown.
   Cut out the 4 triangles.
   Use these triangles to make each figure.
   a) a rectangle
   b) a parallelogram
   c) a trapezoid
   d) a rhombus
   e) an octagon
   Draw each polygon on another piece of dot paper.
   Draw any lines of symmetry.

LESSON 4

5. Which straws could you use to make each polygon? Explain your choice.
   a) an equilateral triangle
   b) a rhombus
   c) an isosceles triangle
   d) a kite
   e) a trapezoid

   Sketch each figure on triangular or square dot paper.

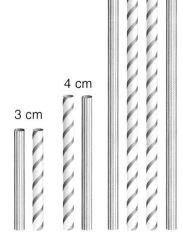

LESSON 5

6. a) The front face of a prism is a rectangle that measures 3 cm by 4 cm.
   Use square dot paper.
   Draw a possible prism.

   b) The front face of a pyramid is an equilateral triangle with side lengths 3 units.
   Use triangular dot paper.
   Draw 2 possible pyramids.

LESSON 6

7. a) How many planes of symmetry does each solid have?

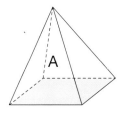 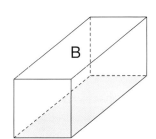

   b) Sketch the face each plane makes where it cuts the prism.

UNIT 3 Learning Goals

- ✓ identify and name polygons
- ✓ construct and classify triangles
- ✓ build and describe figures
- ✓ identify the least information needed to draw a figure
- ✓ draw solids
- ✓ build and describe solids
- ✓ identify planes of symmetry

Unit 3   **105**

Unit Problem: Bridges

You will need:
- Bristol board
- a hole punch or a compass
- paper fasteners
- a centimetre ruler
- centimetre cubes or standard masses

Part 1

Choose one type of bridge truss to build. Your bridge must:
- span a 35-cm gap
- support a load
- stand up by itself

Your teacher will give you a copy of the truss pieces.
Use the truss pieces to cut strips of Bristol board.
How many of each size of strip do you need?
Cut a strip of Bristol board 14 cm wide for the roadway.
How long does the road need to be?
Draw a line 2 cm in from each long edge.
Fold along the lines.

Build the bridge.
How will you brace the top?

Pratt Truss

Double Warren Truss

Part 2

Look at your bridge.
Identify as many of these attributes as you can:
- congruent figures
- scalene, equilateral, and isosceles triangles
- angles that are: right angles; greater than a right angle; or less than a right angle
- equal angles

Name different polygons you see.
Are any of them regular? Explain.

Check List

Your work should show
- [✓] a clear explanation of what you did and why
- [✓] as many attributes as possible
- [✓] how you used what you know about geometry
- [✓] how you found the greatest mass your bridge could support

Part 3

Use two desks or some textbooks to make a 35-cm gap.
Place your bridge across the gap.
Find the load your bridge can support.

Compare your bridge with those of other groups.
Which type of bridge can support the greatest mass?

Write about the bridges and the attributes that make them strong.

Howe Truss

Howe Truss with counter braces

Reflect on the Unit

How can you use what you know about triangles and other polygons to
- combine polygons to make other polygons?
- identify and draw solids?

Cross Strand Investigation

Triangle, Triangle, Triangle

You will need a ruler, several sheets of grid paper, and scissors.

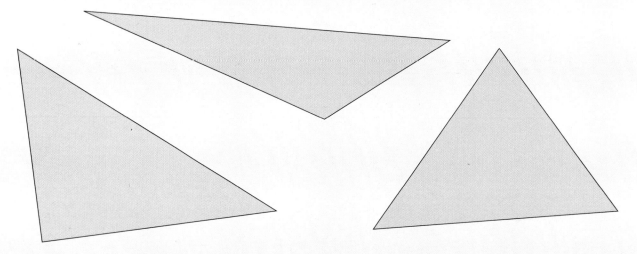

Part 1

➤ On grid paper, draw a large triangle with one angle a right angle. Make sure one side is along a grid line and the third vertex is at a grid point. Estimate the area of the triangle.

➤ On another sheet of grid paper, draw a congruent triangle.

➤ Cut out both triangles. Place the triangles edge to edge to make a rectangle.

➤ Write a multiplication statement to find the area of the rectangle. Calculate the area of the rectangle. Compare the area of the rectangle to the area of the triangle.

Part 2

- Draw a large triangle with each angle less than a right angle. One side should lie along a grid line, with the third vertex at a grid point. Estimate the area of the triangle.
- Draw a congruent triangle.
- Cut out both triangles. Then, cut along a grid line on each triangle to make 2 triangles.
- Arrange the 4 triangles edge to edge to make a rectangle with no gaps or overlaps.
- Write a multiplication statement to find the area of the rectangle. Calculate the area of the original triangle.

Display Your Work

Create a summary of your work. Show all your calculations. Explain your thinking.

Take It Further

- On grid paper, draw a triangle with one angle greater than a right angle. Predict its area.
- How did you use what you learned about the first two triangles to make your prediction?
- Find a way to check your prediction.

UNIT 4

Decimals

Coins Up Close

Learning Goals

- use place value to represent decimals to hundredths
- explore equivalent decimals
- compare and order decimals
- round decimals to the nearest whole number
- estimate decimal sums and differences
- add, subtract, multiply, and divide decimals
- solve problems involving decimals

The coins we use are made at the Royal Canadian Mint in Winnipeg, Manitoba. Coin blanks are punched from large sheets of thin metal. These blanks pass through machines that stamp designs on both sides. The coins are inspected, then loaded into machines that count and package them. The coins are now ready for your pocket or piggy bank.

Key Words

placeholder

equivalent decimals

The Royal Canadian Mint can make 750 coins in a second, or 4 billion coins in a year! To pass inspection, all coins of the same value must have exactly the same mass, thickness, and width. The data below are from the Mint.

Data for Canadian Coins

Coin	Mass (g)	Thickness (mm)	Width (mm)
Penny	2.35	1.45	19.05
Nickel	3.95	1.76	21.2
Dime	1.75	1.22	18.03
Quarter	4.4	1.58	23.88
50¢	6.9	1.95	27.13
Loonie	7	1.75	26.5
Toonie	7.3	1.8	28

The width is 28 mm.

28 mm

- How many different ways can you compare the coins?
- Which coin has the greatest thickness?
- Which coin has a mass of almost 4 g?
- Which coin has a width of almost 3 cm?
- What is the difference in mass between a toonie and a quarter?
- Which 2 coins together have a mass of about 8 g?
- About how many nickels would make 1 kg?
- Write a question you could answer using the data in the table.

LESSON 1

Tenths and Hundredths

The Royal Canadian Mint has been producing pennies, nickels, dimes, quarters, and fifty-cent coins since 1908. Over the years, these coins have changed in mass, thickness, and width.

Explore

You will need Base Ten Blocks.
This table shows how the mass of the penny has changed over the years.

Use Base Ten Blocks to model and compare each mass.
Record your work.

Changes in Mass of the Penny

Dates	Mass (g)
1908–1920	5.67
1920–1979	3.24
1980–1981	2.8
1982–1996	2.5
1997–1999	2.25
2000–2003	2.35

Show and Share

Share your work with another pair of classmates. How do the masses compare?

Connect

➤ The toonie was first produced in 1996. It is 1.8 mm thick.

Here is one way to model 1.8.

Ones	• Tenths	Hundredths

 represents 1.

 represents 0.1.

▪ represents 0.01.

1.8 is a decimal.
It can also be written as the mixed number $1\frac{8}{10}$.
Both are read as "one and eight-tenths."

112 LESSON FOCUS | Explore decimals with tenths and hundredths.

➤ You can also say that the toonie has a thickness of 0.18 cm.
Here is one way to model 0.18.

1.8 mm = 0.18 cm

For a decimal less than 1, you write a 0 in the ones place for 0 ones.
0.18 can be written as a fraction: $\frac{18}{100}$
Both are read as "eighteen-hundredths."

➤ The loonie was first produced in 1987. Loonies made since 1988 have a thickness of 1.75 mm.
Here is one way to model 1.75.

1.75 can be written as the mixed number $1\frac{75}{100}$. Both are read as "one and seventy-five hundredths."

➤ From 1968 to 1999, the dime had a mass of 2.07 g.
Here is one way to model 2.07.

2.07 can be written as the mixed number $2\frac{7}{100}$. Both are read as "two and seven-hundredths."

➤ You can use a place-value chart to show decimals.

Ones	Tenths	Hundredths
1	8	
0	1	8
1	7	5
2	0	7

Practice

Use Base Ten Blocks when they help.

1. Write a decimal and a fraction or mixed number for each picture.

a) [Ones: 2 flats | Tenths: 3 rods | Hundredths: (empty)]

b) [Ones: 4 flats | Tenths: 2 rods | Hundredths: 3 small cubes]

c) [Ones: (empty) | Tenths: 6 rods | Hundredths: 6 small cubes]

d) [Ones: 1 flat | Tenths: (empty) | Hundredths: 9 small cubes]

2. Colour hundredths grids to show each decimal.
Write it as a fraction or a mixed number.
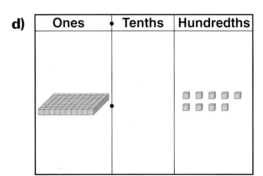
a) 3.17   b) 2.06
c) 0.78   d) 1.4

3. Model each decimal with Base Ten Blocks.
Make a sketch to record your work.
a) 0.46   b) 3.04   c) 1.9   d) 1.09   e) 3.35

4. Sketch a place-value mat.
Show each number on it.
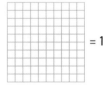
a) 12 hundredths
b) 1 and 27 hundredths
c) 6 tenths
d) 4 hundredths
e) 7 and 89 hundredths
f) 2 and 4 tenths

5. Use Base Ten Blocks to model each fraction or mixed number. Then write it as a decimal.
   a) $\frac{85}{100}$
   b) $4\frac{6}{10}$
   c) $\frac{9}{10}$
   d) $2\frac{8}{100}$
   e) $7\frac{7}{100}$

6. Write each number as a decimal.
   a) forty-six hundredths
   b) five and six-hundredths
   c) two and eight-tenths
   d) nine and forty-nine hundredths

7. Use the data in the table.
   a) Write the number that has a 5 in the hundredths place.
   b) Write the number that has a 3 in the ones place.
   c) Write a number that has a 3 in the hundredths place.

Coin	Width (mm)
Penny	19.05
Dime	18.03
Quarter	23.88
50¢	27.13
Loonie	26.5

8. Draw boxes like these:

   ☐.☐☐ and ☐☐.☐

   Use the digits 4, 5, and 8.
   a) Write these digits in the boxes to make as many different decimals as you can.
   b) Write a mixed number for each of your decimals.
   c) Arrange your numbers to show you have found all possible decimals.

9. The dime is 1.22 mm thick. Sketch Base Ten Blocks to show the decimal 1.22 in as many different ways as you can.

Reflect

What is the purpose of the decimal point in a decimal? Use words and numbers to explain.

Numbers Every Day

Number Strategies

Write each number in expanded form, and in words.

2005
20 500
250 000

ASSESSMENT FOCUS | Question 8

LESSON 2

Equivalent Decimals

Explore

You will need Base Ten Blocks and hundredths grids.
Model each pair of decimals in as many ways as you can.

0.3 and 0.30   0.6 and 0.60

1.8 and 1.80   2.5 and 2.50

Record your work by colouring hundredths grids.

Show and Share

Share your work with another pair of students.
Discuss what you discovered about the pairs of decimals.

Connect

One row of this hundredths grid is one-tenth of the grid.
Each small square is one-hundredth of the grid.

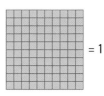 = 1

4 rows are 4 tenths.

0.4

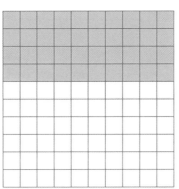

40 squares are 40 hundredths.

0.40

Both 0.4 and 0.40 name the shaded part of the grid.
So, 0.4 = 0.40
Decimals that name the same amount are called **equivalent decimals**.

116 LESSON FOCUS | Identify equivalent decimals.

Practice

1. Write two equivalent decimals that name the shaded part of each grid.

   a) 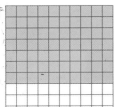   b) 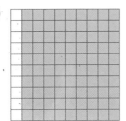   c)    d)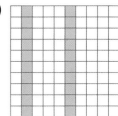

2. Colour hundredths grids to show each number. Write an equivalent decimal.
   a) 0.20   b) 0.9   c) 0.70   d) 0.5

3. Write an equivalent decimal for each number.
   a) 0.5   b) 0.80   c) 0.30   d) 0.6   e) 2.4
   f) 1.70   g) 4.9   h) 7.20   i) 5.50   j) 0.10

4. Find the equivalent decimals in each group.
   a) 0.5   0.05   0.50
   b) 0.70   0.7   0.07
   c) 4.8   4.08   4.80
   d) 6.04   6.40   6.4

5. Which bottle of olive oil is the better buy? Explain how you know.

6. Ruby said that 8.40 is greater than 8.4 because 40 is greater than 4. Was Ruby correct? Use words, pictures, or numbers to explain.

Numbers Every Day

Number Strategies

Estimate each sum.

4567 + 7654 + 6547

1234 + 4321 + 3214

8009 + 9008 + 8090

Which strategies did you use?

Reflect

Use numbers, pictures, or words to explain why 7 tenths is the same as 70 hundredths.

ASSESSMENT FOCUS | Question 6

LESSON 3

Comparing and Ordering Decimals

Olympic medal winners in racing events often beat their competitors by fractions of a second. Very small numbers can make a huge difference!

In the 2002 Winter Olympics, the gold medal-winning bobsled team from Germany beat one American team by 0.3 of a second and another by 0.35 of a second.

Explore

Here are the top 5 results of the 2000 Summer Olympics women's 100-m dash.

➤ Who finished first? Who finished last?
➤ Order the runners from first to last. Use any materials or strategies you wish.
➤ Suppose you do not have enough Base Ten Blocks. How can you order the decimals?

Women's 100-m Dash

Athlete	Time (seconds)
M. Jones (USA)	10.75
T. Lawrence (Jamaica)	11.18
M. Ottey (Jamaica)	11.19
Z. Pintusevych (Ukraine)	11.2
E. Thanou (Greece)	11.12

Show and Share

Share your results with another pair of students. Discuss the strategies you used to order the times.

118 LESSON FOCUS | Compare and order decimals to hundredths.

Connect

Here are the top 3 results of the 2000 Olympics women's pole vault.

Here are 3 ways to find who won gold, silver, and bronze:

Women's Pole Vault

Athlete	Height (m)
S. Dragila (USA)	4.6
V. Flosadottir (Iceland)	4.5
T. Grigorieva (Australia)	4.55

➤ Use Base Ten Blocks to model the height of each pole vault, then put the heights in order.

Ones	Tenths	Hundredths	
(4 flats)	(6 rods)		4.6
(4 flats)	(5 rods)	(5 small cubes)	4.55
(4 flats)	(5 rods)		4.5

➤ Write each decimal in a place-value chart.

Ones	Tenths	Hundredths
4	6	0
4	5	0
4	5	5

Use equivalent decimals.
6 tenths is 60 hundredths.
5 tenths is 50 hundredths.

Look at the whole number parts. All 3 numbers have 4 ones.

Compare the decimal parts. 60 hundredths is the greatest. 50 hundredths is the least.

So, the heights from greatest to least are:
4.6 m, 4.55 m, 4.5 m

Unit 4 Lesson 3

➤ Use a number line.
Mark a dot for each number on the number line.

Since the best result is the greatest height,
read the numbers from right to left: 4.6, 4.55, 4.5

So, Dragila won gold, Grigorieva won silver,
and Flosadottir won bronze.

Practice

Use Base Ten Blocks or a number line to model and compare decimals.

1. Choose the decimal that best describes the amount in each glass.
   a) 0.35 0.60 0.85   b) 0.95 0.38 0.50   c) 0.35 0.10 0.01

2. Copy and complete. Use >, <, or =.
   a) 0.60 ☐ 0.6   b) 0.45 ☐ 0.62   c) 4.07 ☐ 4.12
   d) 5.60 ☐ 5.12   e) 3.08 ☐ 3.8   f) 1.27 ☐ 0.99

3. Write the decimals in order from least to greatest.
   a) 1.47, 1.82, 1.31   b) 2.07, 2.01, 1.85
   c) 0.83, 0.73, 1.04   d) 7.30, 7.62, 6.80

4. Write the decimals in order from greatest to least.
   Think about equivalent decimals when you need to.
   a) 0.5, 0.40, 0.52   b) 16.4, 14.79, 14.09
   c) 0.43, 0.6, 0.55   d) 0.5, 1.4, 1.16

5. Copy. Then write a decimal to make each statement true.
   How did you decide which decimal to write?
   a) 0.45 > ☐   b) 12.7 < ☐   c) 7.01 > ☐
   d) 1.3 < ☐   e) 3.24 > ☐   f) 0.09 < ☐

6. Use the data in the table.
   a) Which sprinter was fastest?
   b) Which sprinter was slowest?
   c) Whose time was faster than Drummond's, but slower than Thompson's?
   d) Look at the number line. Match each letter with the sprinter who made that time.

Men's 100-m Dash

Athlete	Time (seconds)
A. Bolden (Trinidad & Tobago)	9.99
D. Chambers (Great Britain)	10.08
J. Drummond (USA)	10.09
M. Greene (USA)	9.87
O. Thompson (Barbados)	10.04

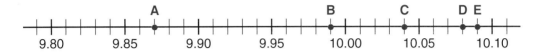

   e) Suppose they raced again.
   Chambers and Greene took 1 second longer.
   The others ran in the same time as in the first race.
   How would this affect the race results?
   Explain.

7. The average Canadian eats 23 kg of beef each year.
   This is about 0.44 kg each week.
   Look at the list.
   a) Who eats less than the weekly average?
   b) Who eats more?

   Beef Eaten Each Week
   Paulette – 0.4 kg
   Luca – 0.43 kg
   Chelsey – 0.49 kg
   Arthur – 0.09 kg
   Holly – 0.52 kg

8. a) Write 3 decimals less than 1.01.
   b) Write 3 decimals greater than 5.8.
   c) Write 3 decimals greater than 4.81 and less than 5.73.
   d) Write your answers to parts a, b, and c in words.

Reflect

How do you know 2.3 is greater than 2.27?
Use pictures, numbers, or words to explain.

Numbers Every Day

Calculator Skills

Find two 5-digit numbers:
- with a sum of 35 428
- with a difference of 35 428

LESSON 4

Rounding Decimals

Suppose there were 5362 spectators at an Olympic swimming event.
Round this number to the nearest ten and nearest hundred.
What strategies did you use?
You can use the same strategies to round decimals.

This table shows the winning times of seven gold medal winners at the 2000 Sydney Olympics.

Event	Athlete	Time (seconds)
Women's 50-m Freestyle Swimming	I. De Bruijn	24.32
Men's 50-m Freestyle Swimming	G. Hall Jr. and A. Ervin	21.98
Women's 100-m Dash	M. Jones	10.75
Men's 100-m Dash	M. Greene	9.87
Women's 100-m Freestyle Swimming	I. De Bruijn	53.83
Men's 100-m Freestyle Swimming	P. van den Hoogenband	48.30

Suppose the times were recorded in whole seconds.
What would each person's time be?
Use any materials you need to help you.

Show and Share

Share your rounded times with a classmate.
Discuss the strategies you used to round.
Why might you want to round the times to the nearest second?

Numbers Every Day

Mental Math

One book costs $10.
How much would each number of books cost?

7

35

128

4327

Connect

➤ In the women's 100-m dash, the gold medal time was 10.75 seconds. The silver medal time was 11.12 seconds.

These times are measured to the nearest hundredth of a second.
To write an estimate of these times, you can *round* to the nearest second.

Here are 2 ways to round 10.75 and 11.12 to the nearest second.

- Use a number line.

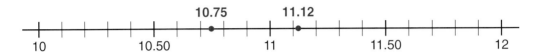

10.75 is between 10 and 11, but closer to 11.
So, 10.75 seconds rounds up to 11 seconds.

11.12 is between 11 and 12, but closer to 11.
So, 11.12 seconds rounds down to 11 seconds.

- Use what you know about decimal hundredths.

Look at the hundredths in 10.75.
You know that 75 hundredths is closer to 1 than to 0.
So, 10.75 rounds up to 11.

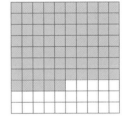

Look at the hundredths in 11.12.
You know that 12 hundredths is closer to 0 than to 1.
So, 11.12 rounds down to 11.

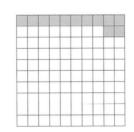

➤ One swimmer had a time of 56.50 seconds in the men's 100-m backstroke.

To round 56.50 seconds to the nearest second:

 Since 50 hundredths is halfway between 0 and 1, you round up.
So, 56.50 seconds rounds to 57 seconds.

Unit 4 Lesson 4 **123**

Practice

1. Round to the nearest whole number.
   a) 4.78   b) 6.31   c) 5.09   d) 1.98   e) 3.2   f) 12.50
   g) 7.49   h) 11.71   i) 40.12   j) 4.47   k) 1.25   l) 3.62

2. Round to the nearest dollar.
   a) $4.78   b) $1.22   c) $7.50   d) $3.99
   e) $6.27   f) $4.49   g) $0.97   h) $21.55

3. Which decimals could be rounded to each circled number? Explain.
   a) ⑦   6.42, 7.1, 6.08   b) ⑨   8.50, 9.43, 9.01
   c) ②   1.9, 2.8, 1.50   d) ④   0.09, 3.89, 4.44

4. The toonie has a mass of 7.3 g and a thickness of 1.8 mm.
   Round each measurement to the nearest whole number.

5. The 1908 penny has a mass of 5.67 g and a width of 25.4 mm.
   Round each measurement to the nearest whole number.

6. The biggest earthworm ever found was 6.7 m long and 2.03 cm wide.
   Round each measurement to the nearest whole number.

7. A runner's time in the 100-m dash is about 11 seconds when rounded to the nearest second.
   Suppose the time was recorded to the nearest hundredth of a second.
   What are the fastest and slowest possible times?

8. Use the digits 4, 5, and 6.
   a) Write as many decimals as you can that round to 5 when rounded to the nearest whole number.
   b) Order the decimals from least to greatest.
   c) Suppose the digit 5 was replaced with 3. How would this affect your results? Explain.

Reflect

Choose a number with a digit in the hundredths place.
How would you round your number to the nearest whole number?

LESSON 5

Estimating Sums and Differences

The 50¢ coin was the first coin produced in Canada.
In recent decades, 50¢ coins have not been widely used.
Nowadays, most of them are purchased by coin collectors.

Explore

Use the data in the table.
➤ Estimate the combined mass of:
  • a 50¢ coin and a quarter
  • a 50¢ coin and a nickel
  • a 50¢ coin and a dime

➤ Estimate the difference in widths of:
  • a 50¢ coin and a penny
  • a 50¢ coin and a quarter
  • a 50¢ coin and a nickel

Coin	Mass (g)	Width (mm)
Penny	2.35	19.05
Nickel	3.95	21.2
Dime	1.75	18.03
Quarter	4.4	23.88
50¢	6.9	27.13
Loonie	7	26.5
Toonie	7.3	28

Show and Share

Compare your estimates with those of another pair of classmates.
Discuss the strategies you used to estimate the sums and differences.

Connect

➤ The toonie has a mass of 7.3 g. The 50¢ coin has a mass of 6.9 g.
To estimate the combined mass of these coins:
Estimate: 7.3 + 6.9

Round each decimal to the nearest whole number.
7.3 rounds to 7.
6.9 rounds to 7.
Add the rounded numbers: 7 + 7 = 14
The combined mass of a toonie and a 50¢ coin is about 14 g.

> You get another estimate if you round just 1 number.
> 7 + 6.9 = 13.9
> So, 7.3 + 6.9 is about 13.9.

LESSON FOCUS | Estimate sums and differences of decimals. **125**

➤ The mass of a penny is 2.35 g.
The mass of a dime is 1.75 g.
To estimate the difference
in these masses:
Estimate: 2.35 − 1.75

Round one or both decimals
to a "friendly number."
Round 2.35 down to 2.25.
2.25 − 1.75 = 0.50
So, 2.35 − 1.75 is about 0.50.

The difference in masses of a penny and a dime is about 0.50 g.

> You get another estimate if you round up to a "friendly number."
> Round 2.35 up to 2.50.
> 2.50 − 1.75 = 0.75
> So, 2.35 − 1.75 is about 0.75.

Practice

1. Estimate each sum.
   a) 4.6 + 9.8
   b) $2.31 + $8.79
   c) 5.99 + 1.40
   d) 11.20 + 6.31
   e) $12.36 + $4.08
   f) 7.1 + 4.2

2. Estimate each difference.
   a) 4.7 − 3.8
   b) 8.07 − 7.91
   c) 10.82 − 6.99
   d) $12.99 − $8.50
   e) 4.04 − 2.96
   f) 12.1 − 8.8

3. Estimate each sum or difference.
   a) 9.75 + 5.5
   b) 5.25 − 1.4
   c) 9.9 + 6.42
   d) 7.1 − 2.96

4. Use the data in the poster.
   a) About how much would 2 adult tickets cost?
   b) About how much would tickets for a 14-year-old and a 9-year-old cost?
   c) About how much more is an adult ticket than a senior ticket?

Numbers Every Day

Number Strategies

Write each pattern rule.

1, 3, 7, 13, 21, 31, 43, …

100, 90, 81, 73, 66, 60, …

Write the next 3 terms in each pattern.

MovieLand
Adult - $13.95
Senior - $8.25
Student - $10.75
Under 13 - $8.50

5. Agnes has $50 in her pocket.
   Explain how Agnes might estimate to check
   if she has enough money to buy the sweater and the jeans.

6. Kenichi estimated that 23.58 + 11.14 is about 35.
   Ariel's estimate was 34.58.
   a) How did each student estimate the sum?
      Use numbers and words to explain.
   b) Use a different strategy to estimate the sum.
      How is your estimate different?

7. Jamel is saving up for a helmet that costs $39.99.
   So far, he has saved $14.10.
   About how much more money does Jamel need to save?

8. An insect's body is 2.34 cm long. Its antennae are 4.58 cm long.
   Estimate the total length of the insect. Show your work.

9. Choose the better estimate. Explain your thinking.
   a) 9.2 − 3.8   4, 5, or 6
   b) 19.6 + 12.2   31, 32, or 33
   c) 10.53 − 4.99   5, 6, or 7

Reflect

Explain how estimating decimal sums
and differences is the same as
estimating whole-number sums
and differences. How is it different?

At Home

Talk with family members.
When do they estimate?
How do they estimate?
Write about what you found out.

LESSON 6

Adding Decimals

Explore

Lindy rides her scooter to school.
Lindy's mass, including her helmet, is 28.75 kg.
The mass of her backpack is 2.18 kg.
➤ About what mass is Lindy's scooter carrying?
➤ Find the total mass the scooter is carrying.
Use any materials you think will help.
Record your work.

Show and Share

Share your results with another pair of classmates.
Discuss the strategies you used to estimate the mass,
and to find the mass.
Were some of the strategies better than others? How?
Explain.

Connect

Julio rides his skateboard to school.
Julio's mass is 26.79 kg.
The mass of his backpack is 2.60 kg.
What total mass is Julio's skateboard carrying?

Add: 26.79 + 2.60
Here are 2 ways to find 26.79 + 2.60.
➤ Use Base Ten Blocks.
   Model 26.79 and 2.60 on a place-value mat.

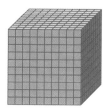

represents 10.

128   LESSON FOCUS | Add decimals to hundredths.

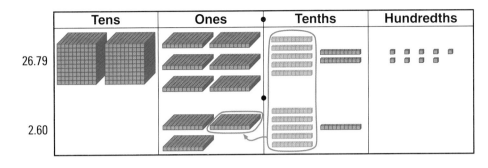

26.79 + 2.60 = 29.39
So, the total mass is 29.39 kg.

➤ Use place value.

Estimate.
26.79 rounds to 27.
2.60 rounds to 3.
27 + 3 = 30

Tens	Ones	Tenths	Hundredths
2	6	7	9
	2	6	0

Step 1: Record the numbers. Align the numbers as you aligned the blocks on the place-value mat.

```
  26.79
+  2.60
```

Step 2: Add as you would with whole numbers.

```
   1
  26.79
+  2.60
  29.39
```

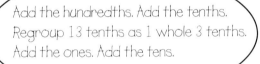

Add the hundredths. Add the tenths. Regroup 13 tenths as 1 whole 3 tenths. Add the ones. Add the tens.

29.39 is close to 30, so the answer is reasonable.

Practice

1. Estimate first. Then add.

   a) 4.6
      + 2.3

   b) 9.5
      + 5.4

   c) $6.25
      + $3.92

   d) 5.24
      + 6.99

Unit 4 Lesson 6 **129**

2. Write each sum vertically, then add.
   a) 27.39 + 48.91   b) 58.09 + 6.40   c) $31.74 + $2.86

3. Add. Think about equivalent decimals when you need to.
   a) 7.56 + 4.8   b) 7.6 + 3.85
   c) 0.3 + 4.71   d) 0.62 + 0.9
   e) 20.48 + 9   f) 10 + 3.7

To add 7.56 + 4.8, I know 4.8 is equivalent to 4.80. So, I write a 0 after 4.8 to show place value.

4. Tagak needed 2.43 m and 2.18 m of rope for his dog team.
   When he added the two lengths, he got the sum 46.1 m.
   Tagak realized he had made a mistake.
   How did Tagak know? What is the correct sum?

5. Lesley bought a CD for $19.95 and a DVD for $26.85.
   How much did she pay for the two items?

6. Paul bought a piece of ribbon 4.9 m long.
   He cut it into 2 pieces. What lengths could the 2 pieces be?
   How many different answers can you find?

7. Hannah bought 2 lobsters. One had a mass of 0.75 kg.
   The other had a mass of 0.9 kg.
   What was the total mass of the lobsters?

8. What is the perimeter of this tile?

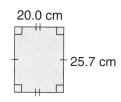

20.0 cm
25.7 cm

9. The perimeter of a rectangle is 74.2 cm.
   What might the dimensions be?
   How do you know?

10. Write a story problem that uses the addition of two decimals with hundredths. Solve your problem. Show your work.

Reflect

Can the sum of 2 decimals with hundredths be a decimal with only tenths?
Use an example to explain.

Numbers Every Day

Calculator Skills

Find two 2-digit numbers with a product of 720.

How many different pairs of numbers can you find?

Make 2!

You will need coloured markers.
Your teacher will give you a set of decimal cards
and hundredths grids.

The object of the game is to shade hundredths grids
to represent a decimal that is as close to 2 as possible.

➤ Shuffle the decimal cards.
   Place the cards face down in a pile.
   Turn over the top 4 cards.
➤ Players take turns choosing one of the 4 cards displayed.
   Each time, the card is replaced with the top card in the deck.
➤ On your turn, represent the decimal on one of the hundredths grids.
   Use a different colour for each decimal.
   You may not represent part of the decimal on one grid
   and the other part on the second grid.
   You may not represent a decimal that would more than fill a grid.
   If each of the decimals on the 4 cards is greater than either decimal
   left on your grids, you lose your turn.
➤ Continue playing until neither player can choose a card.
   Find the sum of the decimals you coloured on your grids.
   The player whose sum is closer to 2 is the winner.

Spinning Decimals

You will need Base Ten Blocks.
Your teacher will give you place-value mats and a spinner.
The object of the game is to make the greatest decimal using the fewest Base Ten Blocks.

Players take turns.

➤ On your turn, you must take rods and unit cubes.
  Spin the pointer 2 times.
  After the first spin, you may choose to take that number of rods or that number of unit cubes.
  After the second spin, take that number of unit cubes or rods, whichever you did not choose the first time.
➤ Make as many trades of Base Ten Blocks as you can.
  Record the decimal for that turn.
➤ After 3 rounds of play, find the sum of your decimals.
  The player with the highest score is the winner.

LESSON 7

Subtracting Decimals

Explore

This chart shows the average annual snowfall in several Canadian cities.

Choose two cities from the chart. Estimate how much more snow one city gets than the other. Then find the difference. Use any materials you think will help. Record your work.

Average Annual Snowfall

City	Snowfall (m)
Regina, SK	1.07
St. John's, NL	3.22
Toronto, ON	1.35
Vancouver, BC	0.55
Yellowknife, NT	1.44

Show and Share

Share your results with another pair of classmates. Discuss the strategies you used to find the difference in snowfalls.

Connect

St. John's, Newfoundland, gets an average of 3.22 m of snow a year.
Halifax, Nova Scotia, gets 2.61 m.
How much more snow does St. John's get than Halifax?

Subtract: 3.22 − 2.61

Here are 2 ways to find 3.22 − 2.61.
➤ Use Base Ten Blocks.
  Model 3.22 on a place-value mat.

  You cannot take 6 tenths from 2 tenths.
  Trade 1 whole for 10 tenths.

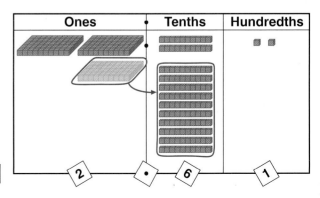

LESSON FOCUS | Subtract decimals to hundredths.

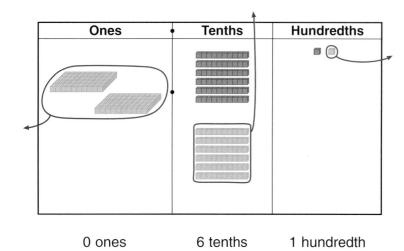

Take away 1 hundredth.
Take away 6 tenths.
Take away 2 ones.
3.22 − 2.61 = 0.61

0 ones   6 tenths   1 hundredth

➤ Use place value.

Estimate.
3.22 rounds to 3.
2.61 rounds to 3.
3 − 3 = 0

Step 1: Record the numbers. Align the numbers to match the blocks and cards on the place-value mat on page 135.

$$3.22$$
$$-2.61$$

Ones	Tenths	Hundredths
3	2	2
2	6	1

Step 2: Subtract as you would with whole numbers.

$$\overset{2\ 12}{\cancel{3}.\cancel{2}2}$$
$$-2.61$$
$$\overline{0.61}$$

When a decimal has no whole number part, you write a zero in the ones place.

Subtract the hundredths. Regroup 1 whole as 10 tenths. Subtract the tenths. Subtract the ones.

Check your answer by adding:

$$\overset{1}{2}.61$$
$$+0.61$$
$$\overline{3.22}$$

The answer 0.61 is close to 0, so the answer is reasonable.
So, St. John's gets 0.61 m more snow than Halifax.

Use Base Ten Blocks to model the decimals.
Write the differences vertically if it helps.

1. Estimate first. Then subtract.
   a) 7.8 – 2.3   b) 6.7 – 3.8   c) 9.35 – 4.26   d) $10.62 – $4.07

2. Subtract.
   a) 6.04 – 3.78   b) 2.76 – 0.98   c) $9.03 – $7.28
   d) 11.09 – 9.29   e) 12.26 – 3.91   f) 73.40 – 54.23

3. Subtract. Think about equivalent decimals when you need to.
   a) 0.56 – 0.4   b) $16 – $4.26   c) 0.8 – 0.36

4. Erin subtracted 12 from 37.8 and got a difference of 36.6.
   How did Erin know she had made a mistake?
   What is the correct answer?

5. Use the data in the table.

   Average Annual Precipitation

City	Precipitation (cm)
Calgary, AB	39.88
Victoria, BC	85.80
Montreal, QC	93.97
Whitehorse, YT	26.90
Winnipeg, MB	50.44

   a) What is the difference in precipitation between Calgary and Whitehorse?
   b) How much more precipitation does Montreal get than Winnipeg?
   c) How much less precipitation does Whitehorse get than Winnipeg?
   d) What is the difference in precipitation between the cities with the greatest and the least precipitation?

6. Use the data in question 5.
   Find which two cities have a difference in precipitation of:
   a) 45.92 cm   b) 8.17 cm   c) 54.09 cm

Math Link

Media

A headline in a newspaper writes a large number like this:

1.5 Million People Affected by Power Cut

We say 1.5 million as "one point five million" or "one and a half million."
1.5 million is one million five hundred thousand, or 1 500 000.

7. In the men's long jump event, Marty jumped 8.26 m in the first trial and 8.55 m in the second trial. What is the difference between his jumps?

8. Candida got a $50 bill for her birthday. She bought a camera for $29.95 and a wallet for $9.29. How much of the $50 is left?

9. Write a story problem that uses the subtraction of two decimals with hundredths.
Trade problems with a classmate.
Solve your classmate's problem.

10. Brad estimated the difference between 11.42 and 1.09 as less than 10. Is Brad correct?
Show 2 different ways to estimate that support your answer.

Reflect

How is subtracting decimals like subtracting whole numbers?
How is it different?
Use words, pictures, or numbers to explain.

Numbers Every Day

Mental Math

Copy and complete each number pattern.

22, 33, □, 55, □, □

24, □, 48, 60, □, □

LESSON 8

Multiplying Decimals by 10 and 100

Explore

You will need a calculator.

➤ Use a calculator to find each product.

2.5 × 10   43.7 × 10   2.76 × 10   14.81 × 10
2.5 × 100   43.7 × 100   2.76 × 100   14.81 × 100

Record the products in a place-value chart.

Thousands	Hundreds	Tens	Ones	Tenths

How can you predict the product when you multiply by 10? By 100?

➤ Find each product. Then check with a calculator.

4.8 × 10   26.8 × 10   3.8 × 10   24.68 × 10
4.8 × 100   26.8 × 100   3.8 × 100   24.68 × 100

Show and Share

Share what you discovered with another pair of classmates.
What patterns do you see?
How can you mentally multiply a decimal by 10?
How can you mentally multiply a decimal by 100?

LESSON FOCUS   Mentally multiply decimals with tenths and hundredths by 10 and 100.

Connect

➤ You can use Base Ten Blocks to multiply.
Multiply: 3.6 × 10
Use Base Ten Blocks.
Model 10 groups of 3.6.

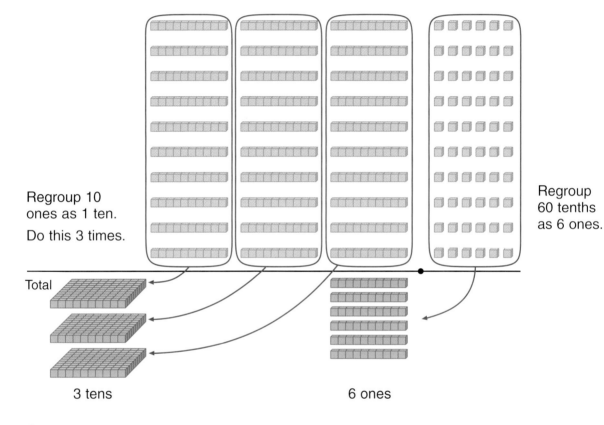

Regroup 10 ones as 1 ten.
Do this 3 times.

Regroup 60 tenths as 6 ones.

3 tens

6 ones

So, 3.6 × 10 = 36

You can use place value to understand what happens when you multiply a decimal by 10 and by 100.

➤ Multiply by 10.
- Multiply: 3.6 × 10
  3.6 = 3 ones and 6 tenths

  3 ones × 10 = 30 ones = 3 tens
  6 tenths × 10 = 60 tenths = 6 ones
  3 tens + 6 ones = 36
  So, 3.6 × 10 = 36

Tens	Ones	Tenths
	3	6

Tens	Ones	Tenths
	3	6
3	6	

- You can use mental math to multiply a decimal by 10.

When you multiply a decimal by 10, the digits shift 1 place to the left.

You can show this by moving the decimal point 1 place to the right.

2.47 × 10 = 24.7

6.8 × 10 = 68

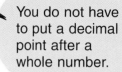

You do not have to put a decimal point after a whole number.

14.81 × 10 = 148.1

➤ Multiply by 100.
- Multiply: 3.6 × 100

    3 ones × 100 = 300 ones
     = 30 tens
     = 3 hundreds
    6 tenths × 100 = 600 tenths
     = 60 ones
     = 6 tens
    3 hundreds + 6 tens = 360
    So, 3.6 × 100 = 360

Hundreds	Tens	Ones	Tenths
		3	6
3	6	0	

- You can use mental math to multiply a decimal by 100.

When you multiply a decimal by 100, the digits shift 2 places to the left.

You can show this by moving the decimal point 2 places to the right.

2.47 × 100 = 247

6.8 × 100 = 680

You write a zero when there are no digits in the ones place.

14.81 × 100 = 1481

Practice

1. Use a place-value chart. Record each product in the chart.
   a) 7.9 × 10   b) 2.67 × 10   c) 0.7 × 100   d) 42.3 × 100

2. Multiply. Use mental math.
   a) 4.7 × 10   b) 62.8 × 10   c) 3.85 × 10   d) 17.45 × 10
      4.7 × 100   62.8 × 100   3.85 × 100   17.45 × 100

3. Use mental math to multiply.
   a) 1.6 × 10   b) 4.82 × 10   c) 53.7 × 10   d) 26.31 × 10
   e) 3.05 × 100   f) 56.73 × 100   g) 0.5 × 10   h) 0.09 × 100

Use mental math to solve each problem.

4. a) Ana bought 10 cans of apple juice.
      How many litres did she buy?
      How much did Ana pay?
   b) Hans bought 100 cans of apple juice.
      How many litres did he buy?
      How much did Hans pay?

5. Fiona read about a man in India who grew
   a bush 18.59 m tall in his garden.
   Fiona said the bush was 185.9 cm tall.
   Was she correct? Explain.

6. A hot dog at the ballpark sells for $2.75.
   The vendor sold 100 hot dogs.
   How much money did she collect?

7. A nickel has a mass of 3.95 g.
   A dime has a mass of 1.75 g.
   Which has the greater mass:
   10 nickels or 20 dimes?

Reflect

Explain how to use mental math
to multiply 8.93 by 10 and by 100.

Numbers Every Day

Number Strategies

Estimate each product.

348 × 9

7125 × 5

25 × 35

Which strategies did you use?

LESSON 9

Dividing Decimals by 10

Explore

You will need a calculator.

➤ Use a calculator to find each quotient. Record the quotients in a place-value chart.
How can you predict the quotient when you divide by 10?

| 15.5 ÷ 10 | 4.6 ÷ 10 | 0.4 ÷ 10 |
| 48.3 ÷ 10 | 9.7 ÷ 10 | 0.1 ÷ 10 |

Hundreds	Tens	Ones	Tenths	Hundredths

➤ Find each quotient.
Then check with a calculator.

10.2 ÷ 10   0.3 ÷ 10   6.1 ÷ 10   9.4 ÷ 10

Show and Share

Share your ideas with another pair of classmates.
How can you mentally divide a decimal by 10?

Connect

➤ There are 4.6 m of ribbon to decorate 10 parcels.
How much ribbon will each parcel get?

Divide: 4.6 ÷ 10
Use Base Ten Blocks.
Model 4.6.

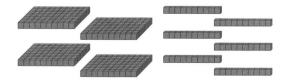

Numbers Every Day

Mental Math

Write an equivalent decimal for each decimal.

0.1
0.20
1.8
25.50

LESSON FOCUS | Mentally divide decimals with tenths by 10.

141

To divide these blocks into 10 equal groups,
first trade each 1 whole for 10 tenths.

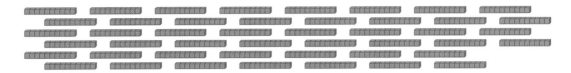

There are 46 tenths.
Divide these into 10 equal groups.
There are 4 tenths in each group, with 6 tenths left over.

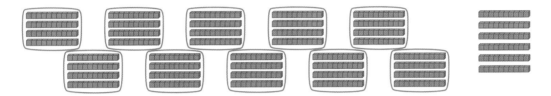

Trade each 1 tenth for 10 hundredths.
There are 60 hundredths.

Divide these 60 hundredths among the 10 equal groups.

There are 4 tenths and 6 hundredths, or 46 hundredths in each group.
So, 4.6 ÷ 10 = 0.46
Each parcel gets 0.46 m of ribbon.

➤ You can use place value to understand what happens
when you divide a decimal by 10.

Divide: 4.6 ÷ 10

4.6 = 4 ones and 6 tenths
    = 46 tenths
    = 460 hundredths
460 hundredths ÷ 10 = 46 hundredths
So, 4.6 ÷ 10 = 0.46

Ones	Tenths	Hundredths
4	6	
0	4	6

➤ You can use mental math to divide a decimal by 10.

$26.5 \div 10 = 2.65$

$427.9 \div 10 = 42.79$

$8.4 \div 10 = 0.84$

$0.7 \div 10 = 0.07$

When you divide a decimal by 10, the digits shift 1 place to the right.

You can show this by moving the decimal point 1 place to the left.

When there are no ones, you use zero as a **placeholder**. Sometimes you need zero as a placeholder in the tenths place.

Practice

1. Use a place-value chart. Record each quotient in the chart.
   a) $8.8 \div 10$
   b) $4.2 \div 10$
   c) $25.1 \div 10$
   d) $16.7 \div 10$

2. Use Base Ten Blocks to divide.
   a) $25.3 \div 10$
   b) $185.3 \div 10$
   c) $8.2 \div 10$
   d) $0.9 \div 10$

Use mental math to solve each problem.

3. Luke jogged 10 laps around the track. He jogged a total of 7.5 km. How far is it around the track?

4. Ten Grade 5 spelling books have a mass of 4.5 kg. What is the mass of 1 spelling book?

5. Jen has 2.8 m of fabric. She will make 10 placemats, all the same size. How long will each mat be?

6. Stanley divided a 2.5-kg bag of dog food equally among his 10 dogs. How much food did each dog get? Show your work.

Reflect

Use Base Ten Blocks to find $14.7 \div 10$.
Explain how this model shows how to divide by 10 mentally.

ASSESSMENT FOCUS | Question 5

LESSON 10

Strategies Toolkit

Explore

Pike Lake is twice as wide as Char Lake.
Char Lake is 2.5 km wider than Perch Lake.
Bass Lake is 10 times as wide as Perch Lake.
Bass Lake is 45.0 km wide.
How wide are Char Lake, Pike Lake, and Perch Lake?

Show *and* Share

Describe the strategy you used to solve this problem.

Connect

Four students have totem poles.
Scannah's pole is 1.3 m shorter than Uta's pole.
Uta's pole is 2.5 m taller than Sta-th's pole.
Yeil's pole is 10 times as tall as Sta-th's pole.
Yeil's pole is 35.0 m tall.
How tall are Scannah's, Uta's, and Sta-th's poles?

Strategies
- Make a table.
- Use a model.
- Draw a diagram.
- Solve a simpler problem.
- Work backward.
- Guess and check.
- Make an organized list.
- Use a pattern.
- Draw a graph.

Understand

What do you know?
- The height of Yeil's pole is 35.0 m.
- You can use that height to find the other heights.

Plan

Think of a strategy to help you solve the problem.
- You can start with the height of Yeil's pole and **work backward**.

144 LESSON FOCUS | Interpret a problem and select an appropriate strategy.

Use the height of Yeil's pole to find the height of Sta-th's pole.
Use the height of Sta-th's pole to find the height of Uta's pole.
Use the height of Uta's pole to find the height of Scannah's pole.
How tall is each pole?

How can you check your answer?
How could you have solved this problem another way?

Practice

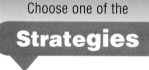

Choose one of the Strategies

1. Karl swam the length of the swimming pool 10 times.
   He swam a total distance of 166.5 m.
   How long is the pool?

2. Lamar opened a 1-L can of tomato juice.
   He poured an equal amount into each of 3 glasses.
   Lamar had 0.4 L left. How much did Lamar pour into each glass?

3. Last weekend, Inga earned money by mowing lawns.
   After her first job, Inga had twice the amount of money she had when she left home.
   After her second job, Inga doubled her money again.
   When Inga got home, she had $18.60.
   How much money did Inga have when she first left home?

Reflect

Choose one of the problems in this lesson.
Use words, pictures, or numbers to explain how you solved it.

Unit 4 — Show What You Know

LESSON

Use Base Ten Blocks when they help.

1. **1.** Write each fraction or mixed number as a decimal.
    a) $\frac{9}{10}$
    b) $3\frac{14}{100}$
    c) $20\frac{1}{100}$
    d) $\frac{67}{100}$

   2. Use the digits 1, 3, 4, 8, and a decimal point.
   Write the number that is closest to 40.

2. **3.** Write an equivalent decimal for each decimal.
    a) 0.8
    b) 7.20
    c) 1.1
    d) 0.60

   4. a) When would you write 7.50 instead of 7.5?
   b) When would you write 7.5 instead of 7.50?

3. **5.** Use the digits 3, 4, 5, and a decimal point.
   Write all the possible decimals.
   Order these decimals from least to greatest.

   6. Use the digits 0, 5, 8 in these boxes: ☐.☐☐
   a) Write the greatest number.
   b) Write the least number.

2/3. **7.** Copy and complete. Use >, <, or =.
   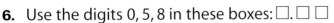
   a) 3.04 ☐ 3.19   b) 0.40 ☐ 0.4   c) 1.7 ☐ 1.25

3. **8.** Write the decimals in order from greatest to least.

   a) 5.62, 5.9, 5.30   b) 0.95, 0.6, 1.3

   9. Copy each statement. Write a decimal to make the statement true.
   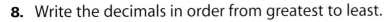
   a) 0.85 > ☐   b) 7.62 < ☐   c) 0.04 > ☐

4. **10.** Round each decimal to the nearest whole number.
    a) 2.34   b) 0.6   c) 0.84   d) 5.5

   11. An athlete's long jump, rounded to the nearest metre, was 8 m.
   What are the longest and shortest lengths the jump
   could have been, in decimals to the hundredths?

146 Unit 4

12. Estimate each sum or difference.
    a) 1.7 + 4.9   b) 7.3 − 2.45   c) 28.1 + 3.14
    d) 12.98 − 4.01   e) 6.78 + 3.12   f) 7.8 − 0.9

13. Add or subtract.
    a) 3.48 + 7.62   b) 14.7 − 8.35   c) 1.98 + 6.3

14. Tran wrote 4 numbers on 4 counters and placed them in a paper bag.
    He drew 2 counters and added the numbers.
    He replaced the counters and repeated the steps 4 more times.
    The sums were: 2.4, 2.5, 2.6, 2.7, 2.8
    Which 4 numbers did Tran write on the counters?

15. Lisa saved $9.26. Her grandpa gave her $4.75 more.
    a) How much money does Lisa have?
    b) How much more money does Lisa need to buy a remote control car that costs $19.95?

16. A canoe is 5.67 m long. How many centimetres is that?

17. A nickel is 21.2 mm wide. Suppose 100 nickels were laid side by side in a line. How long would the line of nickels be?

18. Leonardo lives in a 10-storey apartment building. The building is 121 m tall. All the storeys have the same height. How high is each storey?

19. Use mental math to multiply or divide.
    a) 2.7 × 10   b) 8.46 × 100
    c) 15.8 ÷ 10   d) 32.4 ÷ 10
    e) 52.73 × 10   f) 0.2 ÷ 10

UNIT 4 Learning Goals

- use place value to represent decimals to hundredths
- explore equivalent decimals
- compare and order decimals
- round decimals to the nearest whole number
- estimate decimal sums and differences
- add, subtract, multiply, and divide decimals
- solve problems involving decimals

Unit 4   **147**

Unit Problem: Coins Up Close

 Use a calculator when it helps.

1. Gloria, Lino, and Lisette worked together doing odd jobs for their neighbours. At the end of the day, they had earned exactly 6 of each of the 7 kinds of coins.
   a) How much money did they earn?
   b) The children shared the money equally. How much money did each child get?

2. One toonie has a mass of 7.3 g. Use these data to find the masses of all the numbers of toonies in the table. Copy and complete the table. You may add, subtract, or multiply. Award yourself points for each solution using this point system:
   - 5 points for multiplying
   - 3 points for subtracting
   - 1 point for adding

   What strategy can you use to get the most points? Explain your strategy.

Number of Toonies	Total Mass (g)
1	7.3
2	
3	
4	
5	
6	
7	
8	
9	
10	

3. Could you lift a bag containing $10 000 in toonies? Explain.

Check List

Your work should show
- ☑ the strategies you used to solve each problem
- ☑ how you added, subtracted, multiplied, or divided to find your answers
- ☑ a clear explanation of each answer
- ☑ the correct use of mathematical language and symbols

Coin	Thickness (mm)	Width (mm)
Penny	1.45	19.05
Nickel	1.76	21.2
Dime	1.22	18.03
Quarter	1.58	23.88
50¢	1.95	27.13
Loonie	1.75	26.5
Toonie	1.8	28

4. On the Planet of the Giants, all objects are 10 times as long, wide, and tall as they are on Earth.
Use the table of coin data to answer these questions.
   a) One giant's coin has a width of 265 mm. Which coin is it?
   b) Another giant's coin is 15.8 mm thick. Which coin is it?
   c) How high would a stack of 100 loonies be on the Planet of the Giants? Write your answer in millimetres, centimetres, and decimetres.

5. Make up your own problem about the coins on the Planet of the Giants. Solve your problem.

Reflect on the Unit

Explain why estimating is important when working with decimals.

Unit 4 **149**

Units 1–4 Cumulative Review

UNIT 1

1. The first 2 terms of a pattern are 3 and 5.
   Write 5 different patterns that start with these 2 terms.
   List the first 6 terms for each pattern.
   Write each pattern rule.

2. Copy and complete the table
   for this Input/Output machine.
   Choose 5 input numbers.
   Find each output number.

Input	Output

3. Here is a pattern of figures made with congruent squares.
   The side length of each square is 1 unit.

      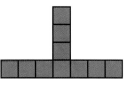

   Frame 1   Frame 2   Frame 3

   a) Find the perimeters of the figures in the first 3 frames.
      Record the frame numbers and the perimeters in a table.
   b) Write the pattern rule for the perimeters.
   c) Use the pattern to predict the perimeter of the figure
      in Frame 12.

UNIT 2

4. Write each number in standard form.
   a) 600 000 + 40 000 + 3000 + 10
   b) five hundred three thousand nine hundred two

5. Find each sum or difference.
   Use mental math when you can.
   a)   7081   b)   4576   c)   8879   d)   1213
      + 1199   − 4149   − 988   + 2208

6. Find each product or quotient.
   a) 4 × 6000   b) 30 × 20   c) 11 × 300   d) 132 ÷ 12
   e) 27 × 68   f) 3576 ÷ 8   g) 74 × 55   h) 9198 ÷ 7

150

7. Caleb has 2691 marbles to share among 9 people.
How many marbles will each person get?

8. Use a ruler.
Measure the sides of each triangle.

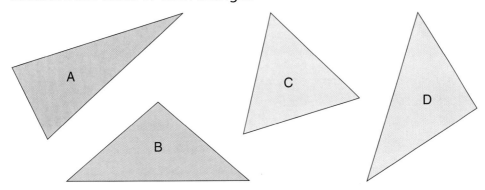

   a) Name each triangle as scalene, equilateral, or isosceles
   b) Which triangles have lines of symmetry?
      How do you know?

9. You will need a triangular pyramid.
Draw the pyramid on dot paper.

10. Order the numbers in each set from least to greatest.
   a) 653 107, 651 370, 635 710, 670 153, 67 531
   b) 3.7, 3.68, 3.86, 3.2
   c) 0.75, 0.8, 0.57, 0.6

11. Write an equivalent decimal for each decimal.
   a) 0.50   b) 3.2   c) 6.70   d) 0.3

12. Find each sum or difference.
   a) 4.3 + 6.8   b) 31.5 − 3.15   c) 26.07 + 3.46   d) 8.3 − 0.68

13. Sophia had $10.47. She spent $4.69.
How much money does she have left?

14. Carter is 183 cm tall.
What is Carter's height in metres?

UNIT 5
Data Analysis
In the Lab

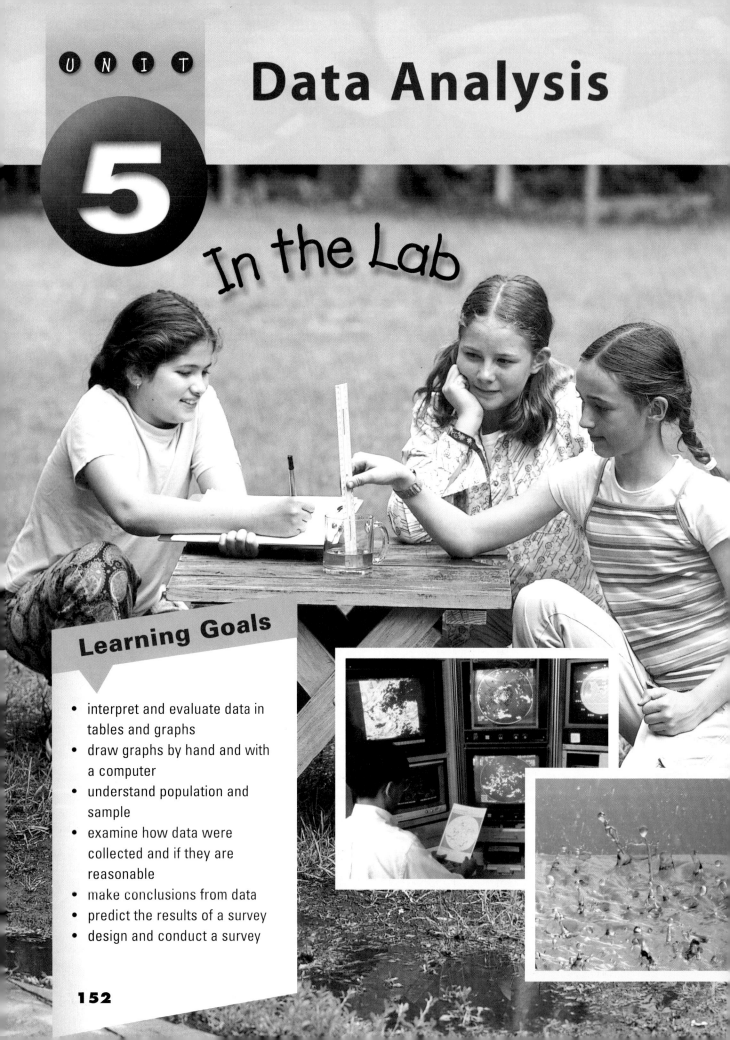

Learning Goals

- interpret and evaluate data in tables and graphs
- draw graphs by hand and with a computer
- understand population and sample
- examine how data were collected and if they are reasonable
- make conclusions from data
- predict the results of a survey
- design and conduct a survey

Key Words

range

line plot

frequency table

frequency

intervals

broken-line graph

population

sample

inference

Students collected data about the amount of rainfall.

Monthly Rainfall for One School Year

Month	Amount (mm)
September	70
October	63
November	67
December	62
January	47
February	46
March	58
April	65
May	67

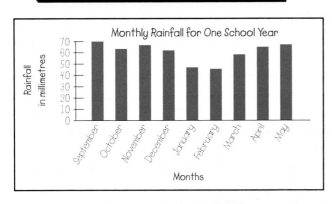

- In which month did the most rain fall?
- In which month did the least rain fall?
- How else could you display the data?
- How do you think these data were collected?
- What other weather data might you record?

153

LESSON 1

Interpreting Data

Governments and organizations collect and use data to help make decisions. Before a decision is made to redesign a park, data are collected on its current use.

Explore

Fun Times Park Saturday Activities

Activity	Number of People
Bicycling	112
Rollerblading	93
Running	35
Riding scooters	51
Skateboarding	43

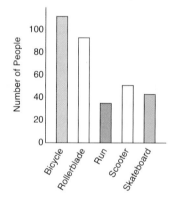

Fun Time Park Saturday Activities

Which activity has the most participants?
Write 5 other things you know from the data.
What do the data *not* show?
Do the table and the bar graph show the same data?
How do you know?
How might the data have been collected?
Do you think the data are reasonable? Explain.

> If we subtract the least value from the greatest, we find the **range** of the data. It tells how spread out the data are.

Show and Share

Write a question that can be answered from the data. Exchange questions with another pair of classmates. Answer your classmates' question.

154   LESSON FOCUS | Read and interpret data in tables and graphs.

Connect

Fun Times Park rents equipment.
The rental data are displayed in graphs.

Tables, pictographs, and bar graphs each show data in an organized way.
The title of the graph tells you what data are displayed.
Each graph presents the same data.

> **Numbers Every Day**
>
> **Number Strategies**
> Write an equivalent decimal for each number.
> - 0.3
> - 1.9
> - 2.60
> - 9.80
> - 6.5

➤ In a pictograph, symbols show the data.
A key shows what each symbol represents. In the pictograph,

🚶 represents 20 rentals.

To find the number of scooters rented, count the 🚶 and multiply by 20.

There are $8\frac{1}{2}$ 🚶.

$8 \times 20 = 160$

🦵 represents 10 rentals.

$160 + 10 = 170$
So, 170 scooters were rented.

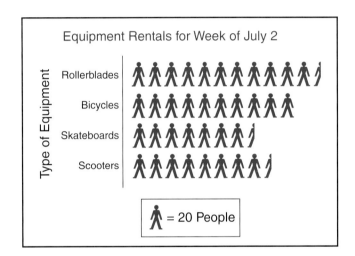

➤ In a bar graph, bars show the data.
The numbers on the vertical axis show the scale.
In this bar graph, the bar for rollerblade rentals is about halfway between 220 and 240.
So, the rollerblade rentals are about 230.
From the bar graph, the greatest number of rentals is 230.
The least number of rentals is about 150.
So, the range is about 230 − 150, or about 80.

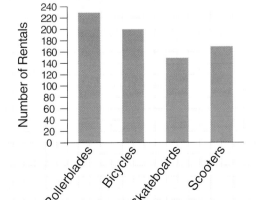

➤ From both the pictograph and bar graph, we can only estimate the number of pieces of equipment rented.
It is usually easier to estimate the number from a bar graph.
We use the scale to do this.
A pictograph has more impact; it is visually appealing.
We do not need a scale to estimate how many.
We know the number to the nearest 10.
For the actual number, we would need to see the table of data used to draw the graphs.

Practice

1. Look at this table of weather data in Canada.
   a) Which city has the most wet days? The fewest?
   b) Which city has the greatest snowfall? The least?
   c) Which cities have about 3 times as much precipitation as Calgary?
   d) Write 2 other questions that can be answered from the data. Answer the questions.
   e) Describe other ways you could display these data.

Average Annual Precipitation in Canadian Cities

City	Snowfall (cm)	Total Precipitation (mm)	Wet Days
Charlottetown, PE	338.7	1201	177
Quebec City, QC	337.0	1208	178
Ottawa, ON	221.5	911	159
Calgary, AB	135.4	399	111
Vancouver, BC	54.9	1167	164
Victoria, BC	46.9	858	153

2. This table shows the after-school activities students chose at Allgood Elementary School.
   a) Which activity was chosen by the most students? The fewest?
   b) Would you use a bar graph or pictograph to display the data? Explain.
   c) Do you think the data would be the same in your school? Explain.

Activity	Number of Students
Choir	18
Math and Science Club	24
Computer Club	36
Fitness Program	60
Homework Club	42

3. This pictograph shows the number of days Members of Parliament (MPs) sat in the House of Commons in one year.
   a) What months are not shown? Why do you think they are not shown?
   b) How many days did MPs sit in the House of Commons in March?
   c) How many days did MPs sit during the year?
   d) Suppose the graph was redrawn with this key.

   🏛 = 8 days

   How would the graph change?

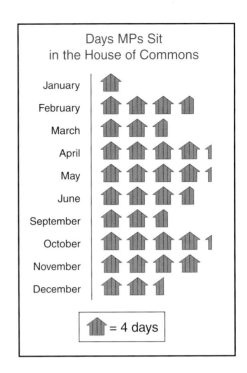

4. This bar graph shows the number of days vegetables grow before they are picked.

   a) Which vegetable takes longest to grow?
   b) What is the range of the data?
   c) Suppose you wanted to display these data as a pictograph. What key would you use? How many symbols would you need for each vegetable?

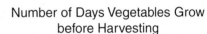
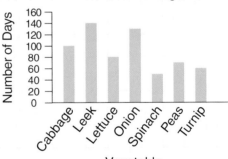

Reflect

Which do you find easier to read: a graph or a table? Explain.

At Home

Look through newspapers and magazines, or search the Internet. Find a table of data you can graph. Graph the data. Give reasons for your choice of graph.

LESSON 2

Frequency Tables and Line Plots

Explore

You will need a paper or Styrofoam cup.
You will conduct an experiment to find out
how a cup lands when it falls.

➤ Slowly slide an upright cup off the edge
of the desk. Record its position after it lands.
➤ Copy and complete this table for 50 results.

➤ Do you think the results would be different
if you rolled the cup off the table? How could you find out?

Show and Share

Share your results with another pair of classmates.
What other ways could you have conducted this experiment?
How could you graph the results?
How often would a cup land upright in 100 rolls off the table? Explain.

Connect

Students in a class timed how long it took to write the alphabet backward.
Here are their results, in seconds.

60, 55, 56, 65, 70, 56, 57, 63, 59, 56, 66, 59, 57, 62, 64,
55, 59, 69, 61, 63, 62, 58, 62, 56, 67, 59, 55, 59, 67, 60

LESSON FOCUS | Make frequency tables and draw line plots.

➤ One way to graph these results is with a **line plot**.
Draw a number line.
Label it with all the numbers in the data.
Make an X above each number, each time it occurs.

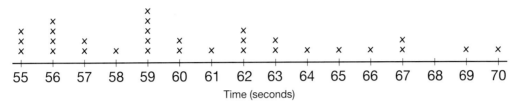

Label the number line.
Write a title for the graph.
From the graph:
- The shortest time was 55 seconds.
- The longest time was 70 seconds.
- Five students took 59 seconds to write the alphabet.

➤ We can organize the data in a **frequency table**.
We group the data in **intervals** of 5 seconds.
We mark a tally for each time in each interval.
We add the tallies to get the total, or **frequency**, in the third column.

Time Interval (seconds)	Tally	Frequency
55-59	卌 卌 卌	15
60-64	卌 IIII	9
65-69	卌	5
70-74	I	1

An equivalent fraction for $\frac{15}{30}$ is $\frac{1}{2}$.
An equivalent fraction for $\frac{6}{30}$ is $\frac{1}{5}$.

From the frequency table:
- Fifteen out of 30, or one-half of the students took less than 1 minute to write the alphabet backward.
- Six out of 30, or one-fifth of the students took 65 seconds or longer to write the alphabet backward.

Unit 5 Lesson 2

The line plot shows individual data.
We can tell how many students wrote the alphabet in a specific time.

The frequency table shows grouped data.
It does not show the time each student took.

Practice

1. Here are the shoe sizes of 24 students.
   8, 7, 4, 5, 6, 5, 4, 3, 5, 6, 6, 4,
   5, 6, 5, 6, 5, 8, 6, 7, 6, 7, 6, 6

   a) Show the data in a frequency table.
   b) Show the data in a line plot.
   c) Which shows the data best: the list, the frequency table, or the line plot? Explain.

2. Work with a partner.
   You will need two number cubes labelled 1 to 6.
   Take turns to roll the two cubes.
   Find the sum of the numbers on the cubes.
   Each student rolls the cubes 25 times.
   a) Record the results in a frequency table.
   b) Which sum occurred most often? Least often? Explain.
   c) How do your results compare with those of another pair of students?

3. Write on the board how you travel to school:
   walk, bus, car, cycle, rollerblade, scooter, skateboard
   Use the class data.
   a) Make a frequency table.
   b) Draw a line plot.
   c) Which is the most popular way to get to school?
      Would this result change if you asked this question at a different time of the year?
      Explain.

4. Work with a partner.
   Conduct the experiment in *Connect*.
   You will need a stopwatch or a watch with a second hand.
   Take turns to time each other while you write the alphabet backward.
   No cheating! Start with Z.
   Write the times on the board.
   Use the class data.

   a) Make a frequency table.
      How did you group the data?
   b) What do you know from looking at the frequency table?
   c) Draw a line plot.
   d) What do you know from looking at the line plot?
   e) Which shows the data better: the frequency table or the line plot? Explain.

5. Students were asked to pick a number between 1 and 10 and write it on a scrap of paper. Here are the results:

   1, 2, 5, 7, 3, 7, 5, 3, 5, 7, 8, 5, 6, 7,
   6, 7, 3, 1, 7, 7, 9, 3, 9, 5, 7, 6, 10, 3

   a) Make a frequency table.
   b) Draw a line plot.
   c) Which number occurred most often?
   d) Was any number not chosen?
   e) To answer parts c and d, which was the easiest to use: the list, the table, or the plot? Explain.
   f) Do you think the results would be the same if you surveyed your class?
      How could you find out?

Numbers Every Day

Mental Math

Estimate each sum or difference.

5.39 + 4.41
1.97 − 0.68
7.63 + 10.01
5.92 − 2.98
6.54 + 1.11

Reflect

When you see a list of data, how do you decide the best way to display it?
Use an example from this lesson in your answer.

Creating Spreadsheets
Using *AppleWorks*

Work with a partner.

The town library tracks the number of books people sign out.

Town Library Sign Out Records

Season	Number of Books
Winter	1488
Spring	1151
Summer	976
Fall	1259

Use *AppleWorks*.
Follow these steps to display these data in a spreadsheet.

1. Open a new spreadsheet in *AppleWorks*. Click:

2. To **enter the data**:
   Click cell A1 to select it.
   Type: Town Library Sign Out Records
   Press Enter.

   Click cell A2 to select it.
   Type: Season
   Press Enter.

   Click cell A3 to select it.
   Type: Winter
   Press Enter.

   Enter the rest of the seasons in cells A4 to A6.

   Click cell B2 to select it.
   Type: Number of Books
   Press Enter.

   Click cell B3 to select it.
   Type the data for Winter: 1488
   Press Enter.

   Enter the data for the rest of the seasons in cells B4 to B6.

3. To **calculate the total number of books**:

   Click cell A7 to select it.
   Type: Total
   Press Enter.

   Click cell B7 to select it.
   Type: =SUM(B3..B6)
   Press Enter.

   You have entered the formula to find a sum.
   The sum of the numbers in cells B3 to B6 will be displayed.
   All formulas must begin with an "=" sign.

4. To **format the spreadsheet**:
   Click cell A1 to select it.

   Click: [Format], then click: [Style ▶]

   Click: [Bold   Ctrl+B]

   Repeat for cells A2, B2, A7, A8, B7, and B8.

5. Save your spreadsheet.

   Click: [File], then click: [Save As...   Shft+Ctrl+S]

   Name your file. Then click: [Save]

6. Print your spreadsheet.

   Click: [File], then click: [Print...   Ctrl+P]

   Click: [OK]

7. Record the name of your spreadsheet.
   You will use it again on pages 167 and 173.

I named my file "Town Library Sign Out Records."

Reflect

What are some advantages to using a spreadsheet?
Explain.

LESSON 3

Drawing Bar Graphs

When you collect data it is important to make sure your data are accurate. Always make sure the numbers or measurements you collect are reasonable. Check any that seem unlikely.

Explore

You will need a measuring tape or a metre stick.
Measure each other's height.
Record the heights on the board.
Look at the class data. Are the results reasonable?
What would make you think a result was not reasonable? Explain.
Order the data for the class.
Draw a bar graph to display the data.

Show and Share

Show your graph to another pair of classmates.
Ask them questions about your graph.
Compare your graph with your classmates' graph.
How are your graphs the same? Different?

Connect

Ms. Lindt teaches math to 25 Grade 5 students. She ordered the marks her students received from the least to the greatest.

She wants to display these data in a bar graph.

54, 55, 58, 61,
63, 65, 66, 67, 68,
71, 72, 73, 73, 74,
75, 76, 77, 78,
79, 81, 86, 87,
88, 91, 95

➤ There were too many pieces of data to graph each mark separately.
  First, she grouped the marks into equal intervals.

164 **LESSON FOCUS** | Draw bar graphs by hand.

➤ She made a frequency table.

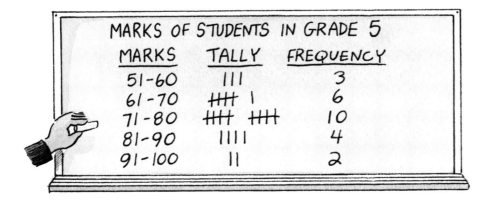

➤ She chose the scale
1 square represents 1 student.
She labelled one axis
"Number of Students"
and the other "Marks."
Then she drew a bar for each
interval of marks and wrote a title.

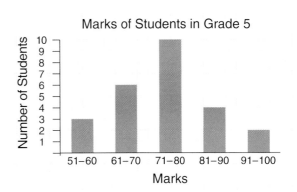

Practice

1. Which of these 3 sets of data would you group into intervals? Why?
   How would you group the data into intervals?
   Draw the bar graph for the data you grouped in intervals.

   a) Election Results for Hillside Public School

Name	Number of Votes
Ho	32
Kake	26
Marr	80
Neigh	30
O'Neil	24
Young	40

   b) Number of Books Read by Students at Flanshaw School

Number of Books	Number of Students
One a week	62
One a month	72
One every 3 months	36
One every 6 months	17
One a year	13

   c) Masses of parcels in a mail room:
   27 kg, 29 kg, 30 kg, 31 kg, 31 kg, 31 kg, 32 kg, 33 kg, 34 kg,
   35 kg, 36 kg, 37 kg, 37 kg, 38 kg, 39 kg, 41 kg, 42 kg

2. Use the table.
   a) What is the range of the data?
   b) Round these data to the nearest 100.
      Then graph the rounded data.
      Which type of graph
      did you choose to draw? Why?
   c) Which province has about double
      the number of police officers as
      British Columbia?
      How does your graph show this?

Number of Police Officers in 2002

Province	Number of Police Officers
Nova Scotia	1 608
Quebec	14 368
Ontario	23 328
Alberta	4 999
British Columbia	7 106

3. A group fitness test showed the number of curls each person could do in 1 minute:
   30, 45, 25, 18, 15, 35, 27, 34, 26, 32, 43, 39, 29,
   31, 43, 44, 26, 16, 20, 40, 44, 22, 27, 30, 36, 37
   a) Choose intervals for the data.
      Make a frequency table.
   b) Choose a suitable graph to display the data.
      Explain your choice.
   c) Draw the graph.

4. The students in a Grade 5 class recorded their heights in centimetres:
   137, 139, 139, 140, 140, 141, 142, 142, 142, 143, 144, 144,
   146, 147, 148, 148, 149, 150, 152, 154, 158, 159, 160
   a) Arrange the data into intervals.
      Create a table to display the data.
   b) Display the data in a bar graph.
   c) Write 2 things you can learn from the graph.

Reflect

Grouping data into intervals makes the data more manageable. Use an example to show why the choice of the interval width is important.

Numbers Every Day

Mental Math

Estimate each product.
Which strategies did you use?
   148 × 9
   211 × 19
   39 × 32
   98 × 102

Drawing Bar Graphs Using *AppleWorks*

Work with a partner.

Use *AppleWorks*.
Follow these steps to graph the Town Library Sign Out data.
Use the spreadsheet you created on pages 162 and 163.

	A	B	C
1	Town Library Sign Out Record		
2	Season	Number of Books	
3	Winter	1488	
4	Spring	1151	
5	Summer	976	
6	Fall	1259	
7	Total	4874	

1. Open a spreadsheet in *AppleWorks*. Click:

2. Open your Town Library spreadsheet.

   Click: |File|, then click: Open... Ctrl+O

   Click on the name of your spreadsheet: Town Library Sign Out Records

   Click: Open

3. To select cells A3 to B6:
   Click cell A3. Hold down the mouse button.
   Drag the cursor from A3 to B6.
   Release the mouse button.

   Follow Steps 4 to 7 to create and print a bar graph of the spreadsheet data.

4. To **create a bar graph**:

   Click: Options , then click: Make Chart... Ctrl+M

   Click the Gallery tab, then click:

   No options should be selected.

LESSON FOCUS | Use a computer to draw bar graphs.

Give the graph a title.
Click the Labels tab. Select Show Title.
Make sure Show Legend is not selected.
Type: Town Library Sign Out Records

Label the axes.
Click the Axes tab. Select X axis.
Type the Axis label: Season
Select Y axis. Type the Axis label:
Number of Books

Enter these settings: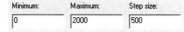

Click: [OK]

5. Save your graph. Click: [File]
   Then click: [Save As... Shft+Ctrl+S]
   Give your file a new name.
   Then click: [Save]

I named my file "Bar Graph."

6. To **move the graph**:
   Move the cursor inside the graph box.
   Click and hold down the mouse button.
   Drag the graph below the total.
   Release the mouse button.

7. To **print the graph**:
   Click: [File], then click: [Print... Ctrl+P]
   Click: [OK]

Reflect

What are some advantages of graphing data using a computer?

LESSON 4

Broken-Line Graphs

Meteorologists record data regularly over time.
In this lesson, you will learn to graph data to show the change in the data.

Explore

Look at this graph.

What does this graph show?
How is it different from other graphs you have seen?
How do the maximum temperatures in May and November compare?
Which months have the same maximum temperature?
Write 4 other questions you can answer from the graph.

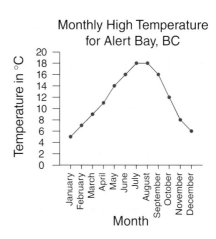

Show and Share

Trade questions with another pair of classmates.
Answer your classmates' questions.
How is this graph the same as a bar graph?
A pictograph? How is it different?

Numbers Every Day

Number Strategies

Order the decimals in each set from least to greatest.
- 0.68, 0.86, 0.80
- 1.35, 5.31, 5.13
- 67.4, 6.74, 7.64
- 2.31, 1.23, 2.13

LESSON FOCUS | Draw broken-line graphs by hand.

Connect

A **broken-line graph** shows data points joined by line segments.

Lucy had a social studies project. She needed to show data of the population of Nova Scotia from 1950 to 1990.

Population of Nova Scotia

Year	Population (thousands)
1950	638
1960	727
1970	782
1980	845
1990	895

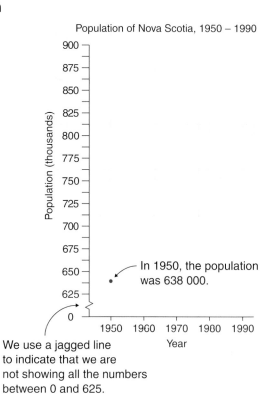

In 1950, the population was 638 000.

We use a jagged line to indicate that we are not showing all the numbers between 0 and 625.

To display these data on a broken-line graph:

➤ Draw two axes.
The horizontal axis shows time.
Label the horizontal axis "Year."
The vertical axis shows the data that change over time.
Label the vertical axis "Population (thousands)."

The range of the data is 257. I don't need to show numbers less than 625. The highest number in the table is 895.

If I start at 625 and count by 25s, the scale will go up to 900.

➤ Choose an appropriate scale.
  Count by 10s for the scale on
  the horizontal axis.
  The horizontal scale is 3 squares
  represent 10 years.
  Count by 25s for the scale on
  the vertical axis.
  The vertical scale is 2 squares
  represent 25.
➤ Mark a point for 1950 at 638.
  Then mark points for the rest
  of the data in the same way.
➤ Use a ruler to connect each
  consecutive pair of points,
  from left to right.
➤ Give the graph a title.
➤ On a broken-line graph, when the line segments:
  • go up to the right, the graph is increasing
  • go down to the right, the graph is decreasing
  The graph goes up to the right.
  The population in Nova Scotia increased from 1950 to 1990.

Practice

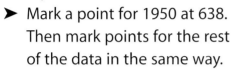

You will need grid paper.

1. A survey of the population of southern sea otters is done each year. The table shows the results from 1998 to 2002.

Year	Number of Otters
1998	1955
1999	1858
2000	2053
2001	1863
2002	1846

   a) Draw a broken-line graph to display these data.
   b) Explain how you chose the vertical scale.
   c) What happened to the number of sea otters from 1998 to 2002? How can you tell from the graph?

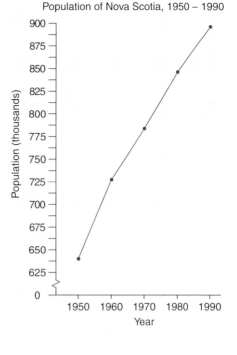

2. This table shows the growth of Rajiv's cucumber vine.

Day	1	2	3	4	5	6	7	8	9	10
Length of Vine (mm)	0	1	7	15	27	35	41	48	53	57

a) Draw a broken-line graph to display these data.
b) What does the line on the graph show?
c) Write 2 things you know from the graph.

3. This table shows the number of beekeepers in Ontario from 1994 to 2003. The numbers are rounded to the nearest 100.

Year	Number of Beekeepers
1994	4500
1995	4300
1996	4100
1997	4100
1998	4000
1999	3600
2000	3000
2001	3000
2002	3000
2003	2700

a) Draw a broken-line graph to display these data.
b) How did you choose the scale?
c) What is happening to the number of beekeepers in Ontario? How does your graph show this?
d) How many beekeepers do you think there will be in Ontario in 2004? Explain your prediction.

Reflect

You can display data using a broken-line graph, a bar graph, or a line plot. Describe a situation that best suits each type of graph. Explain your thinking.

At Home

Look through newspapers and magazines, or on the Internet. Find a broken-line graph. Describe the graph. What information do you get from the graph?

Drawing Broken-Line Graphs Using *AppleWorks*

Work with a partner.

Use *AppleWorks*.
Follow these steps to draw a broken-line graph of the Town Library Sign Out data.
Use the spreadsheet you created on pages 162 and 163.

	A	B	C
1	Town Library Sign Out Record		
2	Season	Number of Books	
3	Winter	1488	
4	Spring	1151	
5	Summer	976	
6	Fall	1259	
7	Total	4874	

1. Open a spreadsheet in *AppleWorks*. Click:

2. Open your Town Library spreadsheet.

   Click: `File`, then click: `Open...  Ctrl+O`

   Click on the name of your spreadsheet: `Town Library Sign Out Records`

   Click: `Open`

3. To select cells A3 to B6:
   Click cell A3. Hold down the mouse.
   Drag the cursor from A3 to B6.
   Release the mouse button.

4. To **create a broken-line graph**:

   Click: `Options`, then click: `Make Chart...  Ctrl+M`

   Click the Gallery tab, then click:

   No options should be selected.

   Give the graph a title.
   Click the Labels tab. Select Show Title.
   Make sure Show Legend is not selected.
   Type: Town Library Sign Out Records

   Label the axes. Click the Axes tab.
   Select X axis.
   Type the Axis label: Season

   Select Y axis.
   Type the Axis label: Number of Books

LESSON FOCUS | Use a computer to draw broken-line graphs.

Enter these settings: Minimum: 0 Maximum: 2000 Step size: 500

Click: OK

5. To **move the graph**:
   Move the cursor inside the graph box.
   Click and hold down the mouse button.
   Drag the graph below the total.
   Release the mouse button.

6. To **print the graph**:

   Click: |File|, then click: Print... Ctrl+P

   Click: OK

7. Save your graph. Click: |File|
   Then click: Save As... Shft+Ctrl+S
   Give your file a new name.

   Then click: Save

I named my file "Broken-Line Graph."

8. Look at the graph.
   Complete each question
   in your notebook.
   a) Describe the shape of the graph.
   b) Write 2 things you can tell from your graph.

9. You drew 2 different graphs to display the same data.
   Look at the 2 graphs.
   a) What can you tell from each graph that you could not tell from the table?
   b) Which graph do you think displays the data better? Why?

Reflect

What data would be best displayed in a bar graph?
A broken-line graph? Explain.

LESSON 5

Interpreting Survey Results

Explore

Suppose you want to find the sport
10- and 11-year-olds most like to watch.
How will you do the survey?
Decide on a survey question.
Predict the result.
Collect data from 8 people in the class.
Make a table to record the results.
How accurate was your prediction?

Show and Share

Share your results with another group of classmates.
Did you ask the same survey question? Explain.
Do you think the results of your survey would be the same
if you surveyed 10 different people in your class?
In another country? Explain.
Do you think the people you surveyed represent all
10- and 11-year-olds? Explain.

Connect

➤ When you collect data to find out something about
all 10- and 11-year-olds, you cannot possibly ask everyone.
All 10- and 11-year-olds are the **population**.
A group of eight 10- and 11-year-olds is a **sample** of the population.

When you want to find out something about all the students
in your school, they are the population.
A sample of the students in your school would be 6 students
from each grade.

LESSON FOCUS | Use a sample to collect data.

175

➤ When you ask a survey question, it must be a **fair** question.
That is, it must not suggest a particular answer.
It must not suggest the answer you predicted.
Suppose you want to find out the favourite type of movie of
the students in your class.

A fair question is, "What is your favourite type of movie?"
The person who answers does not know
your predictions about the answer.

An unfair question is, "Do you agree that comedies
are the most popular type of movie?"
The person who answers knows your opinion
about the question. He might agree to please you.

Practice

1. Write a survey question for each topic.
   a) the favourite food of grade 5 students
   b) the favourite pet of students in your school
   c) the favourite athlete of people in British Columbia

2. Look at each topic in question 2.
   a) What is the population?
   b) What could a sample be?

3. How could you conduct a school survey
   without questioning every student in the school?

Math Link

Your World

Statistics Canada collects data on many topics, such as the economy and the size of the population. Politicians and researchers use these data to learn more about our country and to make decisions.

4. This graph shows the results of a student survey.
   a) Write what the survey question might have been.
   b) How many students do you think were surveyed? Explain.
   c) Do you think a sample or an entire group was surveyed? Why?
   d) Write 2 things you know from this survey.

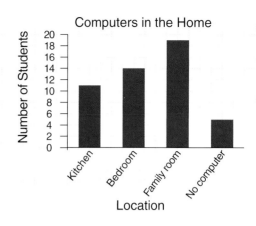

5. You will conduct a survey to find out the favourite breakfast cereal of students in your school.
   a) Will you use a sample or the population in your survey? Explain.
   b) What question will you ask?
   c) Predict the result of the survey.
   d) Conduct the survey and record the data.
   e) Draw a graph to display your data. Explain your choice of graph.
   f) How did the results compare with your prediction?
   g) What else did you find out in your survey?

Reflect

For what survey question could your class be the population?
For what survey question could your class be a sample?

Numbers Every Day

Number Strategies

Round to the nearest dollar.
$ 7.78
$12.62
$ 1.40
$25.35
$14.55

LESSON 6

Making Inferences from Data

Explore

Look at this table.

Sales of Skis						
Month	Sept.	Oct.	Nov.	Dec.	Jan.	Feb.
Pairs of Skis	25	35	40	55	20	10

➤ What happened to the sales of skis from September to December? Why do you think this happened?
➤ What happened to the sales of skis in January and February? Explain.
➤ Will the sales in March be greater or less than the sales in February?
➤ What else do you know from looking at these data?

Show and Share

Share your answers with another pair of students.
If your answers or reasons were different, who is correct?
Could all of you be correct? Explain.

Connect

We can look at data in tables and graphs, and suggest reasons for the results. When we do this, we are making **inferences**.

➤ This table shows the amount of garbage a family put out each week for 6 weeks.

Week	1	2	3	4	5	6
Mass of garbage (kg)	5	5	0	12	4	8

LESSON FOCUS | Using data to make conclusions.

In week 3, there was no garbage.
Perhaps the family forgot to put it out, or the family was on holiday.
In week 4, there was much more garbage.
Perhaps it included the garbage from week 3.
Or the family was working on the house. Or the family had a party.
In week 6, there was more garbage than usual.
The reasons may be similar to week 4.

➤ This broken-line graph shows the depth of water in a dog's bowl for 5 hours.

At 8 o'clock, the depth was 5 cm.
The depth did not change
until 10 o'clock, when it went down to 4 cm.
We infer that the dog drank some water,
or water was spilled from the bowl.
At 11 o'clock, the depth went up to 6 cm.
Someone poured water in the bowl.
At 12 o'clock, the dog drank some water.
At 1 o'clock, the dog drank all the water that remained.

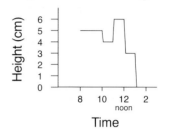

Height of Water in a Dog's Bowl

Practice

1. Best Bicycle Company surveyed 10 bike stores in Manitoba. It wanted to find out how many of their bikes had been sold in the last 6 months. Here is a graph of the data:

a) What is happening to the bike sales?
   Why do you think this is happening?
b) What will happen to the sales in September? Explain.

2. Each of three friends had a bottle of juice to drink at recess. The graphs show the volume of juice in each bottle for 5 hours. Explain how each person drank her or his juice.

   a) Linda b) Jake c) Tara

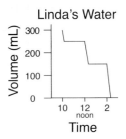
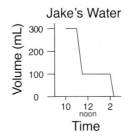
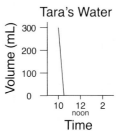

3. The table shows the results of a survey on favourite breakfasts.

   Favourite Breakfast

Cold cereal	8
Hot cereal	3
Eggs	5
Pancakes	17

   What can you infer from the data?

4. This frequency table shows the results of a survey of students aged 10 to 15 years. The question was, "What time do you usually get out of bed at the weekend?"

Time	Tally	Frequency
Before 7 a.m.	|||| |||	8
Between 7 a.m. and 8 a.m.	||	2
Between 8 a.m. and 9 a.m.	||||	4
Between 9 a.m. and 10 a.m.	|||| ||	7
After 10 a.m.	|||| ||||	9

   a) Why might students get out of bed before 7 a.m.?
   b) Christine looked at the table and stated, "Most students don't get out of bed until after 10 a.m." Is she correct? Explain.

5. Five adults recorded the distances they drove in one week.
   All distances were related to driving to and from work, and driving during working hours.

   Suggest reasons for the distances travelled by each person.

Name	Distance (km)
Mr Antak	3
Mr Jordan	12
Ms Ellis	50
Ms Nystrom	400
Ms Davis	2000

6. This table shows the results of a survey to find out the number of television sets in homes.

Number of TVs	0	1	2	3	4	5
Number of homes	2	4	12	11	5	1

   a) Graph the data.
   b) How many homes were surveyed?
   c) Suggest reasons why a home might have 5 television sets.
   d) Do you think the results would be the same in a different part of the country? A different country? Explain.

7. Three students competed in a paper airplane contest. This table shows the distances flown in each of 3 trials.

Student	Dan	Kyla	Sara
Trial 1	124 cm	112 cm	118 cm
Trial 2	84 cm	105 cm	111 cm
Trial 3	103 cm	108 cm	81 cm

   Here are 3 ways to choose a winner.
   Method A: the longest distance on a single flight
   Method B: the total distance of the three flights
   Method C: the best two of the three flights

   a) Who would win for each method?
   b) Whom do you think should win? Explain.

Numbers Every Day

Mental Math

Write 5 different number sentences with the answer 375.

Reflect

How can you make inferences from a table or a graph? Use an example from this lesson to explain.

LESSON 7

Strategies Toolkit

Explore

Greg was playing marbles with his friends.
In the first game, he lost 2 marbles.
In the second game, he lost twice as many as in the first game.
In the last game, he won 8 marbles.
Greg finished with 25 marbles.
How many marbles did he start with?

Show and Share

Describe the strategy you used to solve the problem.

Connect

At school, Jasmin bought chocolate milk for $0.75, a hot dog for $1.00, and a bottle of water for $1.07. On the way home, she found a quarter on the sidewalk. At the end of the day, she had $1.37 in her wallet. How much money did Jasmin start with?

Strategies

- Make a table.
- Use a model.
- Draw a diagram.
- Solve a simpler problem.
- **Work backward.**
- Guess and check.
- Make an organized list.
- Use a pattern.
- Draw a graph.

What do you know?
- Jasmin finished with $1.37.
- Jasmin spent $0.75, $1.00, and $1.07.
- Jasmin found $0.25.

Think of a strategy to help you solve the problem.
- You can **work backward**.
- Start with $1.37.
- Subtract what she found.
- Add what she spent.

182 LESSON FOCUS | Interpret a problem and select an appropriate strategy.

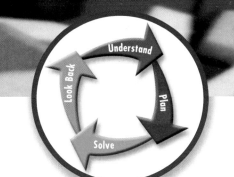

- Subtract $0.25 from $1.37.
- Add $1.07, $0.75, and $1.00.
- How much money did Jasmin start with?

How can you check your answer?
How could you have solved this problem another way?

Practice

Choose one of the **Strategies**

1. On Saturday Jo walked her dog for 20 minutes more than she did on Sunday. For both days she walked a total of 1 hour 15 minutes.
   For how long did Jo walk her dog on Saturday?

2. All 5 students in Carlo's group saved some money each week toward the cost of a trip. The total amount of money saved one week by the group was $12.50.
   Suppose each student saved a different amount.
   How much might each have saved that week?

3. Ari lines up his hockey cards with the same number of cards in each row. The card in the middle of the array has 5 cards above, below, to the right, and to the left.
   How many cards does Ari have?

Reflect

How can working backward help you solve a problem?
Use words and numbers to explain.

Unit 5 Lesson 7 **183**

Unit 5 Show What You Know

LESSON

1

1. This table shows the number of games the top 10 National Basketball Asssociation scorers played in 2003.
   a) Who played the most games? The fewest games?
   b) How many more games did Pierce play than O'Neal?
   c) Write 2 other questions you could answer using these data. Answer the questions.
   d) How else could you display these data?

Name	Games Played
Allen	76
Bryant	82
Duncan	81
Garnett	82
Iverson	82
McGrady	75
Nowitzki	80
O'Neal	67
Pierce	79
Webber	67

2

2. This line plot shows the numbers of games won by students playing Tic-Tac-Toe.
   a) How many students won 7 games?
   b) What was the most common number of wins?
   c) Write 2 other questions you could answer using these data. Answer the questions.

Tic-Tac-Toe Games Won

```
                    x
                    x
            x x x x
            x x x x
            x x x x x
x x x x x x x x     x
1 2 3 4 5 6 7 8 9 10
    Number of Games
```

3. Draw a line plot.
   Here are the masses, to the nearest kilogram, of a group of Grade 5 students:
   26, 28, 30, 32, 32, 32, 33, 33, 34, 35, 35, 36, 37, 37, 37, 37, 38, 38, 39, 39, 40, 41, 42, 43, 43

2
3

4. Use the data in question 3.
   a) Arrange the data into intervals. How did you decide on the intervals?
   b) Make a frequency table.
   c) Draw a bar graph.
   d) Write 3 questions you can answer using the graph. Answer your questions.

184 Unit 5

5. A survey of Canadians aged 9 to 14 was conducted. The table shows how many dollars each spends on music CDs for every $100 spent.
   a) Draw a broken-line graph to show these data.
   b) Describe the graph.
   c) What do you think the data will be for 2003? Explain.

Year	Amount
1999	$19
2000	$22
2001	$17
2002	$16

6. Write a survey question for each topic below. What is the population? What could a sample be?
   a) the favourite shoe style of grade 5 students
   b) the most common colour of car in your neighbourhood

7. Conduct a class survey about favourite hockey teams.
   a) Write the question you will ask.
   b) Conduct the survey.
   c) Display the results.
   d) Write about your survey.

8. This graph shows how the volume of juice in a glass changed in 6 minutes. Describe what you think happened to the juice.

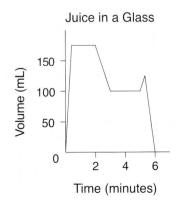

Unit 5 Learning Goals

- ✓ interpret and evaluate data in tables and graphs
- ✓ draw graphs by hand and with a computer
- ✓ understand population and sample
- ✓ examine how data were collected and if they are reasonable
- ✓ make conclusions from data
- ✓ predict the results of a survey
- ✓ design and conduct a survey

Unit Problem

In the Lab

When scientists conduct experiments they collect and study data. You will conduct an experiment to collect and study data.

Part 1

Work in a group.
Decide on a question that may be answered by doing an experiment.
Design the experiment.
What do you expect the results might be?

Part 2

Conduct the experiment.
Remember to:
- Write down what you want to find out.
- List each step of the experiment.
- List any materials you need.
- Record your data in a table.
- Draw and label a graph to display your data.
- Write the results of the experiment.

Check List

Your work should show
- ☑ that you created a plan to answer your experiment question
- ☑ how you collected and recorded data accurately
- ☑ a graph that is easy to understand, with labels and title
- ☑ a clear explanation of your results

	M	T	W	T	F	S	S
WEEK 1	14	16	17	18			
WEEK 2							
WEEK 3							
WEEK 4							

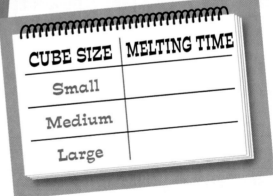

CUBE SIZE	MELTING TIME
Small	
Medium	
Large	

Part 3

Trade results with another group.
Check each other's work.
Are the results reasonable? Explain.
What can you infer from your results?

Part 4

Present your results to the class.
Discuss what you learned from your experiment.

Reflect on the Unit

Describe some ways you can display data.
When would you use each way?
Use examples to explain.

UNIT 6

Measurement

All Aboard!

Learning Goals

- tell time to the nearest second
- read and write time using the 24-hour clock
- explore the relationship between time and distance
- count money and make change
- estimate, measure, and compare mass, volume, and capacity
- relate volume and capacity

DEPARTURES
HALIFAX, NS   12:45 P.M.
TRURO   4:15 P.M.
MONCTON, NB   7:40 P.M.
BATHURST   8:16 P.M.
MATAPÉDIA, QC 10:00 P.M.
DRUMMONDVILLE 6:25 A.M.
MONTRÉAL   8:00 A.M.
KINGSTON, ON 12:31 P.M.

Key Words

seconds
SI notation
24-hour clock
speed
capacity
volume
cubic centimetre (cm³)
displacement
milligrams
tonnes

Aislinn has come to the model train show.

- How is time measured in these pictures?
- How may Aislinn spend her money at the show?
- What objects can a model train boxcar carry? A real boxcar carry?
- How do the masses of the objects compare?
- How could you measure the capacity of a boxcar?

LESSON 1

Measuring Time

It takes about one **second** to clap your hands together one time.

Explore

You will need a stopwatch.

➤ List 6 to 8 fitness activities you can do in the classroom.
➤ Estimate and record how long the first activity will take.
➤ Take turns to do each activity. Record your estimate first. Your partner will time you to the nearest second. Record your work in a table.

That's 48 seconds.

Activity	Estimated Time	Actual Time
Balancing on one foot	60 seconds	2 minutes 17 seconds
20 sit-ups	80 seconds	48 seconds

Show and Share

Discuss your strategies for estimating time. How close were your estimates to the actual times? Did your strategy for estimating change as you did each activity? Explain.

Numbers Every Day

Number Strategies

Order the numbers in each set from least to greatest.

- 8.98, 9.89, 0.99, 0.89, 0.98
- 12.78, 12.87, 71.28, 17.82, 71.82
- 0.08, 1.01, 1.81, 0.8, 0.41

LESSON FOCUS | Tell time to the nearest second.

Connect

Many analog clocks show time in hours, minutes, and seconds.

➤ This clock shows a time of
4 hours 32 minutes 10 seconds.
We write: 4 h 32 min 10 s
In **SI notation**, we write: 04:32:10

04:32:10

➤ One second is a very small unit of time.
There are 60 s in 1 min.
It takes 60 s for the second hand
to move all the way around the clock.

60 s = 1 min

Often, we do not need to tell time
to the nearest second.
We can round time to the nearest minute.

There are 30 s in half a minute.

When there are less than 30 s on the clock, round time back to the full minute.

When there are 30 s or more on the clock, round time forward to the next full minute.

About 09:57

About 05:18

➤ The circled date is October 24th, 2006.
October is the 10th month of the year.
In SI notation, we write: 2006 10 24

2006 10 24

Practice

1. Is each time estimate reasonable? Explain.
   a) Hop on one foot: 45 s
   b) Run all the way around the schoolyard: 2 min 20 s

2. Write each date in SI notation.
   a) June 23, 1966   b) your birthday   c) the first day of next month

3. Write each date in words.
   a) 1994 07 16   b) 2000 01 01   c) 2004 11 09   d) 2026 03 31

4. Look at each clock.
   In SI notation, write the exact time and the time to the nearest minute.
   a)    b)    c)

5. Jack estimated it would take 50 s to jog to his friend's house.

   When he left, his watch looked like this: When he arrived, his watch looked like this:

   a) How long did it take Jack to jog to his friend's house?
   b) Was Jack's estimate reasonable? Explain.
   c) Jack knocked at the door.
      He waited 30 s before someone answered.
      What did his watch look like then? Draw it.
   d) His friend was not home. So, Jack jogged home.
      How long do you think it took him? Explain.

Reflect

Tell about a day when knowing time to the nearest second was important. Tell about another day when it was not important to know the time to the nearest second.

LESSON 2

The 24-Hour Clock

How many seconds are in 1 minute?
How many minutes are in 1 hour?
How many hours are in 1 day?

Explore

Here are flight departure times from Vancouver, and their destinations.

01:15 Hong Kong	14:55 London
06:00 Toronto	20:50 London
11:10 Toronto	23:45 Toronto
15:15 Hong Kong	

➤ Which flights leave very early in the morning? How do you know?

➤ Which flights leave before noon?

➤ Which flights leave after noon? How do you know?

➤ Which flights leave late at night? How do you know?

Show and Share

Share your results with another pair of students.
Compare answers.
If the answers do not agree, try to find out who is correct.
Why do you think the times are written in this way
instead of using a.m. and p.m.?

LESSON FOCUS | Read and write time on a 24-h clock.

Connect

If your friend says she will be at your house at 8 o'clock, you need to know if she means 8 a.m. or 8 p.m. There is another way to write the time where we do not use a.m. or p.m. We use a **24-hour clock**.

➤ There are 24 h in one day.
From midnight to noon, the hours are from 0 to 12.
From 1 o'clock to midnight, the hours are from 13 to 24.
When we use the 24-h clock, we use 4 digits to write the time.

9:45 a.m. is written 09:45

6:15 p.m. is written 18:15

We use a similar notation for time in seconds.

11:14:12 is 14 min 12 s after 11 a.m.

15:27:34 is 27 min 34 s after 3 p.m.

➤ Kathy arrived at the library at 11:45 and left at 14:20.
How long did she spend in the library?
Count on to find the time.

11:45 to 12:00 is 15 min.

12:00 to 14:00 is 2 h.

14:00 to 14:20 is 20 min.

Total time: 15 min + 2 h + 20 min
= 2 h 35 min

Practice

1. Which clocks show the same time?

   a)
   p.m.

   b)

   c)

   d)

   e)
   p.m.

   f)
   a.m.

2. Write each time using a.m. or p.m.

   a)

   b)

   c)

   d)

3. Write each time using a 24-h clock.

   a)
   p.m.

   b)
   p.m.

   c)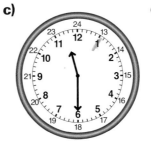
   p.m.

   d)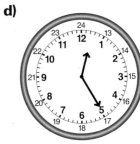
   a.m.

Unit 6 Lesson 2 **195**

4. Here are the times for the first three runners in a race:
   Andrew 2:36:24; Brad 2:35:12; James 2:36:04
   a) Who won the race? Who came second?
      Who came third?
   b) What was the difference in times between the first- and second-place finishers?

5. a) A ferry leaves Port Hardy at 07:30 and arrives in Prince Rupert at 22:45.
      How long is the journey?
   b) An overnight ferry leaves Shearwater at 23:45 and arrives at Port Hardy at 08:10.
      How long is the journey?

6. A bus leaves Anatown at 11:50 a.m. and arrives in Beaconsfield 3 h 25 min later.
   What time does the bus arrive in Beaconsfield?
   Show the time as many different ways as you can.

7. Cat's time for the marathon was 4:23:36.
   The race started at 10 a.m.
   What time did Cat finish her run?
   Write the time in two ways.

Reflect

When you write or tell a time, which way do you prefer: using a 12-h clock or a 24-h clock?
Give examples in your answer.

Numbers Every Day

Mental Math

Estimate each difference.
Which strategies did you use?

4048 − 53
7782 − 3078
1212 − 161
9999 − 9901

LESSON 3

Exploring Time and Distance

Explore

You will need:
- books to raise a ramp
- a marble
- Bristol board, 30 cm by 10 cm
- a metre stick
- masking tape
- a stopwatch

➤ Create a 10-cm high ramp. Use the Bristol board, books, and masking tape as shown. Mark distances of 2 m, 4 m, 6 m, and 8 m from the end of the ramp.

➤ One student lets the marble go at the top of the ramp. Another student measures the time the marble takes to travel 2 m. A third student records the time and distance in a table.

➤ Change roles. Repeat the steps. Measure the times for the marble to travel 4 m, 6 m, and 8 m.

➤ Look at the data in your table. What patterns do you see? Estimate the time the marble would take to travel each distance: 10 m, 12 m, 14 m

Marble Rolling

Time (s)	Distance (cm)

Show and Share

Compare your results with those of another group. Describe your strategy to estimate the times for the marble to travel 10 m, 12 m, and 14 m. How is the distance the marble travels related to the time it takes?

LESSON FOCUS | Explore time and distance relationships.

Speed is a measure of how fast an object is moving.

➤ Helen's marble travels 50 cm every second.
   At this speed, it will go:
   50 cm in 1 s
   100 cm in 2 s
   150 cm in 3 s
   200 cm in 4 s, and so on.

   We can show this:
   • In a table   • On a graph

Time (s)	Distance (cm)
1	50
2	100
3	150
4	200

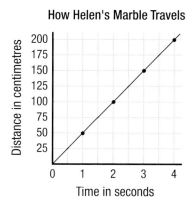

How Helen's Marble Travels

➤ Hannah's marble travels 60 cm every second.
   Hannah's marble goes farther each second than Helen's marble.

My marble moves faster than Helen's marble.

➤ Colin's marble travels 45 cm every second.
   Colin's marble does not go as far each second as Hannah's marble or Helen's marble.

My marble moves slower than Hannah's marble and Helen's marble.

Practice

1. A toy car travels 5 cm every second.
   At this speed, how far will it go in 1 s? 2 s? 3 s? 10 s?

2. A bowling alley is 25 m long.
   The bowling ball travels 5 m in 1 s and 10 m in 2 s.
   The speed of the ball does not change.
   When will the ball hit the pins? Show your work.

3. A car travels 80 km in 1 h and 160 km in 2 h.
   Suppose the car continues at the same speed.
   How long will it take to travel 400 km? Explain.

4. Campbell walks 80 m every minute.
   Tyler walks 85 m every minute.
   a) How far will each person walk in 1 min?
      5 min? 20 min? 1 h?
   b) Who is walking faster? How do you know?

5. a) When has Car A travelled 12 cm?
   b) When has Car B travelled 12 cm?
   c) Which car is moving faster? How do you know?

6. Alyssa canoes 0.75 m every second.
   Jillian canoes 0.72 m every second.
   Which girl canoes slower? How do you know?

7. The Cheung family plans to drive
   500 km to a vacation resort.
   How long do you estimate the trip will take?
   Use pictures, numbers, and words
   to show your thinking.

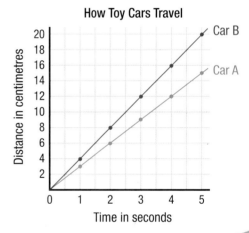

How Toy Cars Travel

Reflect

Suppose you ride your bicycle down the road.
Describe how the time you take and the distance
you travel are related.

Numbers Every Day

Mental Math

Find the next number in each pattern.
Write each pattern rule.
- 1.2, 1.4, 1.6, 1.8, ☐
- 9000, 900, 90, 9, ☐
- 0.05, 0.5, 5, ☐
- 0.2, 0.9, 1.6, ☐

Assessment Focus | Question 7

Unit 6 Lesson 3

LESSON 4

Strategies Toolkit

Explore

Kelly and Ryan leave their home in the forest.
Ryan leaves at 3:00 p.m. He walks 3 km every hour.
Kelly leaves at 4:00 p.m. She walks 4 km every hour.
Kelly follows Ryan's path.
When does Kelly overtake Ryan?
How far has each student walked at that time?

Show *and* Share

Describe the strategy you used to solve this problem.

Connect

David and Pat leave Oshawa to cycle to Oakville.
Pat leaves at 9:00 a.m.
She cycles 15 km every hour.
David leaves at 10:00 a.m.
He cycles 20 km every hour.
When does David overtake Pat?
How far has each person travelled at that time?

Strategies
- Make a table.
- Use a model.
- Draw a diagram.
- Solve a simpler problem.
- Work backward.
- Guess and check.
- Make an organized list.
- Use a pattern.
- **Draw a graph.**

What do you know?
- Pat leaves at 9:00 a.m.
  She cycles 15 km every hour.
- David leaves at 10:00 a.m.
  He cycles 20 km every hour.

Think of a strategy to help you solve the problem.
- You could **draw a graph**.
- On the same grid, draw a line graph for each person.
- Find where the graphs meet.

200 LESSON FOCUS | Interpret a problem and select an appropriate strategy.

Find how far Pat has gone at 10:00 a.m., 11:00 a.m., noon, and so on.
Record the times and distances in a table.
Mark a point on the graph for each time and distance.
Draw a line through the points.
Repeat this strategy for David.
Where do Pat and David meet?
How far has each person gone?
What time is it then?

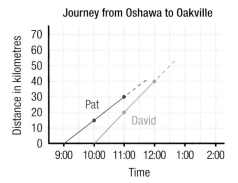

How could you have solved this problem another way?

Practice

Choose one of the **Strategies**

1. A freight train travels 100 km every hour.
   It passes over a river at 10:00 a.m.
   Four hours later, a passenger train passes over the same river.
   The passenger train travels 150 km every hour.
   When does the passenger train overtake the freight train?
   How far are they from the river?

2. Erica leaves Deer Lake to travel 300 km to Gander.
   She drives 80 km every hour.
   Gerry leaves Deer Lake for Gander one hour later.
   He drives 90 km every hour.
   Who reaches Gander first? How do you know?

Reflect

When can you use the strategy of draw a graph to solve a problem?
Use an example to explain.

LESSON 5

Estimating and Counting Money

Explore

How many different ways can you show $1000?

You will need play money and a calculator. Take turns to find different ways to show $1000.
➤ One student finds a way to show $1000.
➤ Another student records the numbers and values of the bills in a table.
➤ A third student checks the total is $1000.
➤ Change roles each turn.

Show and Share

Compare your results with those of another group. What ways did you find that are the same? Different? What else do you notice?

Connect

Here is a collection of money.

202   LESSON FOCUS | Estimate and count money to $1000.

Here's how Raisa counts the money.
➤ First, she sorts the bills and coins into groups.

| 4 x $100 | 2 x $50 | 8 x $20 | $10 | $5 | $2 | $1 | 25¢ 10¢ 5¢ 3x1¢ |
| $400 | $100 | $160 | $10 | $5 | $2 | $1 | 43¢ |

➤ Then, she counts the bills and coins.
➤ Next, she adds the values of the bills and coins:
$400 + $100 + $160 + $10 + $5 + $2 + $1 + $0.43 = $678.43

Here's how Daniel counts the money.
➤ First, he sorts the bills and coins into piles that total $100.

$100 $100 $100 $100 $100 $100 $78.43

➤ Then, he counts: "1, 2, 3, 4, 5, 6 hundred dollars"
➤ Next, he adds on the value of the remaining bills and coins:
$600 + $78.43 = $678.43

Practice

Use play money when it helps.
1. Estimate, then count. Record each estimate and amount.

   a)    b)

2. Make each amount using the fewest bills and coins.
   Record the numbers of bills and coins in a table.
   a) $75.50   b) $166.13   c) $542.86   d) $989.48

3. Callum had a $50 bill, a $20 bill, a $10 bill, and two $5 bills.
   He used 3 bills to buy earrings for his mother.
   Show all the possible combinations of bills he may have used.
   What is the least price for the earrings? The greatest price?

4. Abby counts the coins in her piggy bank.
   The total is $83.77.
   What is the fewest number of coins she could have?
   Show your work.

5. Show three different ways to make $871.56.
   Use pictures, numbers, or words to record each way.

6. Three people have a $100 bill to share.
   How can they share the money?

7. Skye sold some old toys at a yard sale.
   Her little brother helped her wash the toys and set up.
   Skye promised that for every $10 she earned,
   she would give her brother $1.
   a) Look at the shoebox.
      About how much money do you think
      Skye made at the yard sale?
   b) About how much does Skye owe her brother?
      Explain.
   c) Skye wants to know exactly how much money
      she made. How could she count the money?

8. A man leaves for the mall with $355 in bills and coins in his pocket.
   When he reaches in his pocket, he finds he only has $290.
   He knows he is missing 4 bills.
   Which bills and coins might he have lost?
   Show at least three possible answers.

Reflect

When do you need to know the exact
amount of money?
When is an estimate good enough?

Numbers Every Day

Calculator Skills

Find three numbers with a product of 15 600.

Find three odd numbers with a sum of 345.

LESSON 6

Making Change

The Auction Game

The goal is to spend the least amount and buy the most fish and decorations for your fish tank.
You will need assorted play money and a set of auction cards.

➤ Each player starts with $100 and 3 auction cards dealt face down.
➤ Players take turns being the auctioneer.
➤ The auctioneer turns over a card for other players to bid on.
   The highest bidder buys the item.
   The auctioneer makes change and keeps the money from the sale.
   Any card not sold is kept by its auctioneer.
➤ Play continues around the table until each card has been turned over and bid on.

Show and Share

Discuss the strategies you used to bid for items.
What strategies did you use to make change?
How did you know the change you got was correct?

LESSON FOCUS | Make purchases and make change from $100.

Connect

Dakota bought a hamster habitat for $76.83.
He paid for it with three $20 bills, one $10 bill, and two $5 bills.

Here's how the clerk could make change for him:
"That's $76.83 …

$76.84 $76.85 $76.90 $77.00 $78.00 $80.00

That's $3.17 in change."

That's $3.17 in change.

But when the clerk opens her cash drawer,
she finds she has no loonies.

She trades a $20 bill and a $5 bill for a roll of 25 loonies.

Then the clerk makes change for Dakota as shown above.

Dakota gets $3.17 in change.
Dakota estimates to check he got the correct change.

   The hamster habitat cost about $77.
I got about $3 in change.
$77 + $3 is $80.
That's right.

25 loonies are equal to one $20 bill and one $5 bill.

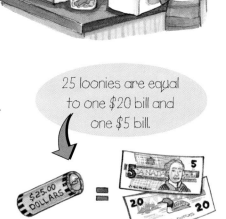

Practice

Use play money when it helps.
1. Draw pictures to show the change for each purchase.
   a) Ailsa bought a pig.
      She paid with two $20 bills and two $10 bills.
   b) Kolby bought a gecko.
      She paid with four $20 bills.
   c) Adam bought a cockatoo.
      He paid with four $20 bills, three toonies, three loonies, and two quarters.

2. Mr. Kwan's class adopted a rabbit from the local animal shelter.
   They bought these supplies.
   The students paid for the items with five $20 bills.
   They were given two bills and
   eight coins as change.
   What bills and coins were they given?

3. Zoe's grandma bought a pair of lovebirds
   for $98.26. She was given $2 change.
   a) How much money did Zoe's grandma
      give the clerk?
   b) Why do you think she gave this amount?

4. Ms. Allen has $100 to spend on dog food.
   Each bag of food costs $32.79.
   a) How many bags of dog food can she buy?
   b) How much change will she get?

5. Michel is saving to buy an aquarium. It costs $82.27.
   Michel has this money.

   a) Does Michel have enough money?
      Do you need to count it all to find out? Explain.
   b) Which bills and coins could Michel use to pay
      for the aquarium? What is his change?
   c) How could Michel pay for the aquarium
      in a different way? What change
      would he get? Show your work.

Numbers Every Day

Number Strategies

Estimate each sum.

0.54 + 1.72

5.69 + 2.48

12.33 + 9.11

9.48 + 2.13

Reflect

When you make a purchase, how can
you check you get the correct change?
Use an example to explain your strategy.

LESSON 7

Capacity

Explore

You will need some containers, measuring spoons, a graduated cylinder, and water.

➤ Predict the order of the containers, from least capacity to greatest capacity.
➤ Measure the capacity of each container.
➤ Order the containers from least capacity to greatest capacity.
➤ How does the order compare with your prediction?

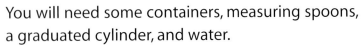

Show and Share

Show your work to another pair of students.
Describe the strategy you used to predict the order of the containers.
How did you measure the capacity of each container?

Connect

Which container has the greatest capacity? The least capacity?

> 1000 mL = 1 L

1.35 L

2.2 L

775 mL

- Container A has capacity 1.35 L.
  This is 1.35 × 1000 mL = 1350 mL
- Container B has capacity 2.2 L.
  This is 2.2 × 1000 mL = 2200 mL
- Container C has capacity 775 mL.

The greatest capacity is 2200 mL. This is container B.
The least capacity is 775 mL. This is container C.

208 LESSON FOCUS | Estimate, measure, and compare capacities in millilitres and litres.

Practice

You will need some containers, a graduated cylinder, and water.

1. Choose three different containers.
   a) Estimate and record the capacity of each container.
   b) Measure and record each capacity.
   c) How do the estimates and measures compare?

2. Order these capacities from least to greatest.
   a) 3.45 L   b) 2120 mL   c) 1.75 L

3. Write each capacity in millilitres.
   a) 5.25 L   b) 2.5 L   c) 4.38 L

4. Choose one container as a benchmark.
   a) Predict how you would sort the other containers into three groups: greater capacity, lesser capacity, and equal capacity
   b) Describe how you could check your sorting. Show your work.

5. Choose a small container and a large container.
   a) Measure the capacity of each container. Fill the large container with water.
   b) Predict the number of times you can fill the small container using the water from the large container.
   c) Find how many times you can fill the small container using the water from the large container.
   d) How do your results compare with your prediction?

Reflect

Describe how you would compare the capacities of two containers.
Use words and pictures to explain.

Calculator Skills

Use the digits 5, 6, 7, 8, and 9.
Find the product closest to 55 000.

☐ ☐ ☐
× ☐ ☐

ASSESSMENT FOCUS | Question 4

LESSON 8

Volume

How could you find out how much space this shoebox takes up?

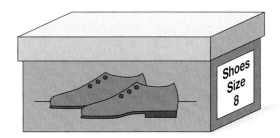

Explore

You will need interlocking cubes.

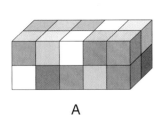   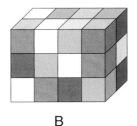   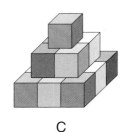

     A   B   C

➤ Estimate how many cubes you need to build each solid.
➤ Build each solid.
  Record the number of cubes you used each time.
➤ Build your own solid. Label it D.
  Find the number of cubes needed to build your solid.
➤ Order the solids, from the one that takes up the most space to the one that takes up the least space.

Show and Share

Share your work with another pair of students.
How did you order the solids?

210   **LESSON FOCUS** | Use centimetre cubes to measure volume.

Connect

The **volume** of an object is a measure of the space it takes up.

A centimetre cube has a volume of one **cubic centimetre** (1 **cm³**).

We can use centimetre cubes to measure volume.

➤ This rectangular prism has 4 rows of 5 cubes, or 20 cubes.
  The volume of this prism is 20 cubic centimetres, or 20 cm³.

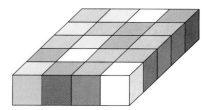

➤ This object has 8 cubes in the bottom layer and 3 cubes in the top layer.
  The volume of this solid is 11 cubic centimetres, or 11 cm³.

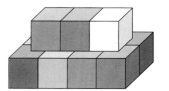

Practice

You will need centimetre cubes.

1. Each prism is made with centimetre cubes.
   Find the volume of each prism.
   Order the prisms from least to greatest volume.

   a)   b)   c)

   d) 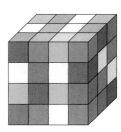   e) 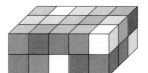   f)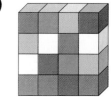

Unit 6 Lesson 8

2. Estimate the volume of each solid.
   Find each volume.

   a) 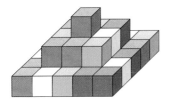   b)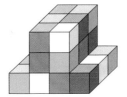

   c) 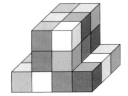   d)

3. Look at the solids in question 2.
   Order the solids from least to greatest volume.

 4. How many different prisms can you make using
   18 centimetre cubes?
   How do you know you have found all of them?
   What is the volume of each prism?

5. Use centimetre cubes.
   Build a solid with a volume of 20 cm^3.
   Your solid should *not* be a rectangular prism.
   Compare your solid with that of a classmate.
   How are the solids the same? How are they different?

6. Describe a strategy you could use to estimate,
   then find the volume of this textbook.
   What problems might you have finding
   the volume? Compare your strategy
   with that of a classmate.

Numbers Every Day

Number Strategies

Write each number in expanded form.
What does the digit 3 represent in each number?

- 76 032
- 38 545
- 90 306
- 53 114

Use centimetre cubes to make a solid.
How would you find the volume of your solid
in cubic centimetres?
Use words and pictures to explain.

LESSON 9

Relating Capacity and Volume

The capacity of a graduated cylinder is 500 mL.
This is how much liquid it can hold.
If we pour in 400 mL of water, we can say the volume of water is 400 mL.
That is, we can measure the volume of water in millilitres.

You will investigate how millilitres are related to cubic centimetres.

Explore

You will need centimetre cubes, a 500-mL graduated cylinder, and water.

➤ Pour 400 mL of water into a
   500-mL graduated cylinder.
   Record the volume of water in a table.
   Place 10 cubes in the cylinder.
   Record the number of cubes added
   and the new volume, in millilitres.
   Calculate and record the change in volume.
➤ Add 10 more cubes.
   Record the new volume.
   Continue to add groups of 10 cubes.
   Each time, record the volume and the
   change in volume.
➤ Describe any patterns you see in the table.
➤ Look at your results.
   When you added 10 cubes, how did
   the volume in the cylinder change?
   How many millilitres equal 10 cm^3?

Number of Cubes Added	Volume (mL)	Change in Volume (mL)
0	400	—
10		

Show and Share

Share the patterns you found with another group of students.
How could you use water in a graduated cylinder to find the volume of a stone?

LESSON FOCUS | Explore the relationship between millilitres and cubic centimetres.

Connect

The volume of an object can be measured in cubic centimetres or millilitres.

$1 \text{ cm}^3 = 1 \text{ mL}$

➤ Here is another way to find the volume of an object.
You can use **displacement** of water to find the volume of this triangular prism.

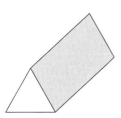

I can't use unit cubes to find the volume of this prism.

| Mark the water level in a container. | ➡ | Totally submerge the prism. Mark the new water level. |

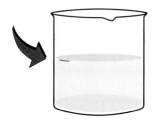

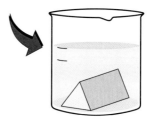

| Remove the prism. Fill the container to the upper mark. Record the volume of water added, in millilitres. | ➡ | Convert the volume in millilitres to cubic centimetres. |

I added 15 mL of water, so the volume of the triangular prism is 15 cm³.

The volume of the triangular prism is 15 cm³.

You will need a container, water, and a graduated cylinder.

1. Collect 4 small solid objects.
   a) Estimate the volume of each object.
   b) Find each volume.
   c) Order the objects from least to greatest volume.

2. Use modelling clay to build a solid.
   Try to make a solid with a volume of 250 cm³.
   a) Find the volume of your solid.
   b) How close is the volume to 250 cm³?

3. Choose two different solids from the classroom.
   Look for solids with about the same volume.
   a) Explain why you chose the solids you did.
   b) Find the volume of each solid in cubic centimetres.

4. a) What is the volume of 100 centimetre cubes?
   b) Put 100 centimetre cubes into an empty graduated cylinder.
      Read the number of millilitres from the scale.
   c) Compare your answers to parts a and b.
      Explain any differences.

5. You will need 50 counters.
   a) Predict the volume of 50 counters in cubic centimetres.
   b) Find the volume of 50 counters.
   c) How does your estimate compare to the volume?

6. Describe how you could find each measure.
   a) the volume of one dime in cubic centimetres
   b) the volume of a toy car in millilitres

Numbers Every Day

Mental Math

Find the missing numbers.
Write each pattern rule.
- 625, 600, ☐, ☐, ☐, 500, 475
- 870, 940, ☐, ☐, ☐, 1220, 1290
- 1250, 1175, ☐, ☐, ☐, 875, 800
- 72, ☐, ☐, 108, ☐, 132

Reflect

Explain how you can use displacement of water to measure the volume of an object.

ASSESSMENT FOCUS Question 3

LESSON 10

Measuring Mass

Explore

You will need balance scales and a set of standard masses.

Find an object that has a mass:
- less than 1 g
- between 50 g and 100 g
- approximately 500 g
- approximately 1000 g, or 1 kg

Find:
- 3 objects with a total mass of about 2000 g, or 2 kg
- 2 objects whose masses differ by 500 g

Record your work in a table.

Show and Share

Which masses was it easy to find objects for? Which were difficult? Why? How did you estimate the mass of each object?

Object	Estimated Mass	Actual Mass
Chocolate milk	500 g	

Math Link

Social Studies

The next time you measure an object, look to see where it was made. Many countries use the metric system to record mass, capacity, and volume. One kilogram is the same in Canada, Singapore, or any other country.

216 LESSON FOCUS | Measure and compare mass in milligrams, grams, and kilograms.

Connect

We use **milligrams** (**mg**), grams (g), and kilograms (kg) to measure mass.

There are 1000 g in 1 kg.

1000 g = 1 kg

➤ An object with a large mass is measured in kilograms (kg).

18 kg

8 kg

1000 mg = 1 g

There are 1000 mg in 1 g.

➤ An object with a small mass is measured in grams (g).

➤ An object with a very small mass is measured in milligrams (mg).

58 g   500 g

Practice

1. Match each object below with its approximate mass. Explain your choices.
   10 g, 2 kg, 600 g, 6 kg, 50 kg, 10 mg

   a)

   b)

   c)

   d)

   e)

   f)

Unit 6 Lesson 10

2. Record the best estimate. Explain your choices.
   a) 100 g, 500 g, 1 kg, or 5 kg

   b) 300 mg, 3 g, 3 kg, or 30 kg

3. a) One mouse has a mass of about 30 g.
      What is the mass of each animal?
      • 1 grasshopper
      • 1 gerbil
      • 1 yellow-collared macaw
   b) About how many mice would balance the macaw?
   c) Suppose you had 1 kg of grasshoppers.
      About how many grasshoppers would you have?
      Explain.

4. The vitamin C tablets in a bottle have
   a total mass of 250 g.
   Each tablet has a mass of 500 mg.
   How many tablets are in the bottle?
   Show your work.

5. Together, Sarah's two cats have
   the same mass as her dog, 11 kg.
   The mass of one cat is 1500 g greater than
   the mass of the other.
   What are the masses of the two cats?

6. Peter eats peanut butter and jelly sandwiches every day for lunch.
   He estimates that he uses 30 g of jelly and 40 g of peanut butter
   per sandwich. How many 1-kg containers of each
   would he need to make his lunches for 40 weeks?
   Use words and numbers to explain.

Reflect

How do you know when to use milligrams,
grams, or kilograms to measure mass?
Give examples in your answer.

Numbers Every Day

Calculator Skills

Find two 5-digit numbers
whose sum is 48 700 and
whose difference is
20 000.

LESSON 11

Exploring Large Masses

Explore

You will need metric bathroom scales.
➤ Choose an object with a large mass.
  Measure its mass.
➤ Choose an object that will not fit on the scales.
  How can you measure its mass?
  Try out your ideas.
  How many of this object would you need to make a 100-kg mass?
  A 1000-kg mass?
  Record your work.

Show and Share

How did you measure the mass of the object that was too large for the scales?
Compare one of your objects and its mass with that of another group.
Which group needed more objects to make 1000 kg?

Connect

The **tonne** (**t**) is a unit of mass.
The tonne is used to measure an object with a very large mass.

There are 1000 kg in 1 t.

The mass of a compact car is about 1 t.
The mass of an elephant is about 4 t.

1000 kg = 1 t

LESSON FOCUS | Use tonnes to measure mass.

➤ The mass of an 18-wheel truck is about 44 t. How many kilograms is that?

$1 \text{ t} = 1000 \text{ kg}$
So, $44 \text{ t} = 44 \times 1000 \text{ kg}$
$= 44\,000 \text{ kg}$
The mass of the truck is about 44 000 kg.

➤ The mass of a walrus is about 1300 kg. Find the mass of a walrus in tonnes.

$1 \text{ t} = 1000 \text{ kg}$
So, $1300 \text{ kg} = \frac{1300}{1000} \text{ t}$
$= 1.3 \text{ t}$
The mass of a walrus is about 1.3 t.

Practice

1. Match each object with its approximate mass. Explain your choices.
   30 kg, 2 t, 400 g, 20 t

   a)    b)    c)    d)

2. Find each mass in kilograms.
   a) The mass of a male giraffe is about 1.9 t.
   b) Debbie needs 0.75 t of concrete mix for her new patio.
   c) A rhinoceros has a mass of about 2.3 t.

3. Find each mass in tonnes.
   a) A garbage truck has a mass of about 16 500 kg.
   b) The mass of a tent trailer is about 600 kg.
   c) The mass of a newborn blue whale is about 2500 kg.

Numbers Every Day

Number Strategies

Estimate the total amount.

$4.69
$1.89
$6.32
+ $4.98

4. Record the best estimate. Explain your choice.
   a) 200 g, 2 t, 50 kg, or 5 kg
   b) 15 g, 2 kg, 10 t, or 500 kg

5. A small truck carrying computer parts has a mass of 8 t.
   The computer parts have a mass of 3500 kg.
   What is the mass of the empty truck? Show your work.

6. One sheet of paper has a mass of about 5 g.
   There are:
   • 500 sheets of paper per package
   • 10 packages per box
   a) What is the mass of 1 package of paper? 2 packages?
   b) How many boxes of paper have a mass of 1 t?
   Use words, numbers, or pictures to explain.

7. Two tonnes of oranges are put into 5-kg bags.
   How many bags are needed?
   Show your work.

8. How many of your math textbooks do you need to make a mass of 1 t?
   Show your work.

At Home

Reflect

How are grams, kilograms, and tonnes related?
When is it better to record a mass in kilograms? In tonnes?

List five objects in or around your home whose mass can be measured in milligrams, grams, kilograms, or tonnes. Which was the most difficult to find? Why?

ASSESSMENT FOCUS | Question 6

Unit 6   Show What You Know

LESSON

1

1. For each clock, use SI notation to:
   - write the exact time
   - write the time to the nearest minute

   a)
   b)

2. Write each date using SI notation.
   a) May 22nd, 2010
   b) January 8th, 2014

3. Vanitia did 50 sit-ups.
   Mark timed her with a stopwatch.
   When Vanitia started, the classroom clock looked like this:
   When Vanitia finished, the stopwatch looked like this:

   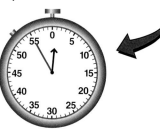

   What time did the classroom clock show when Vanitia finished?

2

4. Write each time using a 24-h clock.
   a) 9:43 a.m.
   b) 7:21 p.m.

3

5. A cyclist travels 20 km in 1 h and 40 km in 2 h.
   Suppose the cyclist continues at the same speed.
   How long will it take her to travel 120 km? Explain.

5

6. What bills and coins can you use to make $657.84?
   Show three different ways.
   Use pictures, numbers, or words to record each way.

LESSON

5   **7.** Find the fewest bills and coins that make each amount.
       a) $9.99   **b)** $99.99   **c)** $999.99
       What pattern do you notice? Explain how the pattern works.

6   **8.** Ray bought some clothes. The total bill was $84.27.
       He paid with three $20 bills and three $10 bills.
       He got a $5 bill and 3 quarters in change.
       Did Ray get the correct change? How do you know?

7   **9.** Order these capacities from greatest to least.
       2.5 L   2150 mL   4 L   1980 mL

8   **10.** Each object is made with centimetre cubes.
       Find the volume of each object.
       Which object has the lesser volume?
       a) 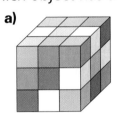   **b)**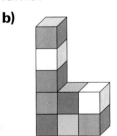

9   **11.** How could you find the volume of a golf ball?
       Use pictures and words to explain.

10
11   **12.** Copy and complete.
       a) 3200 kg = ☐ t   **b)** 21 kg = ☐ g
       c) 8 t = ☐ kg   **d)** 4000 mg = ☐ g
       e) 12 500 kg = ☐ t   **f)** 2 t = ☐ kg

11   **13.** An empty trailer has a mass of 850 kg.
       When the trailer carries a load, its mass is 1.4 t.
       What is the mass of the load?

UNIT 6 Learning Goals

- ✓ tell time to the nearest second
- ✓ read and write time using the 24-h clock
- ✓ explore the relationship between time and distance
- ✓ count money and make change
- ✓ estimate, measure, and compare mass, volume, and capacity
- ✓ relate volume and capacity

Unit Problem

All Aboard!

You will need:
- drinking straws
- flexible drinking straws
- scissors
- Bristol board
- pencil crayons or markers
- a glue stick
- tape
- a calculator
- centimetre cubes

Part 1

Design the track for a model railway.
Use drinking straws to build the track.
Cut the straws if you need to.

You have a budget of $100.
Straight sections of track cost $0.10 per centimetre.
There are extra charges of:
- $0.20 for each bend in the track
- $1.25 for each bridge or overpass

Part 2

As you place each section of track, record your work in a table.
How much change do you have from $100?

Length of Track (cm)	Cost of Track ($)	Extra Charges ($)	Total Cost ($)

Part 3

Suppose the train travels these speeds:
- 3 cm every second on straight sections of track
- 2 cm every second on curved sections of track and bridges

Estimate how long it will take the train to travel once around your track. Calculate the time.

Part 4

Use centimetre cubes to build a model boxcar. Find the volume of your boxcar.

Suppose one real boxcar holds 70 000 kg of cargo. How many real boxcars would you need to carry 120 t of cargo? Explain.

Part 5

Make up your own problem about your model railway.
Solve your problem.
Present your design and results to the class.

Check List

Your work should show
- ☑ a track design that costs less than $100
- ☑ how you decided which pieces of track to use
- ☑ your measurements and calculations recorded correctly, including units
- ☑ how you calculated your answers

Reflect on the Unit

Why do we measure?
Write about the different ways you have learned to measure objects.

UNIT 7

Transformational Geometry in Art

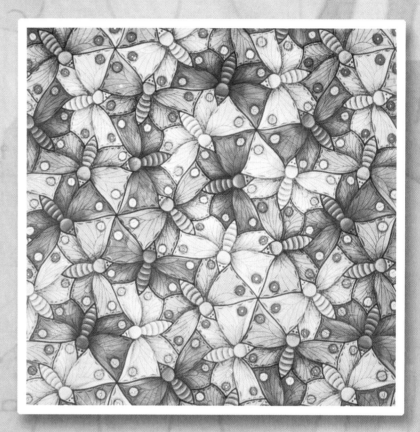

M.C. Escher was a Dutch graphic artist.
He made many drawings like these.

Learning Goals

- recognize a slide (translation), a turn (rotation), and a flip (reflection)
- construct figures with one line of symmetry
- explore tiling patterns and tessellations
- plot and identify points on a coordinate grid

Geometry

Key Words

- translation (slide)
- translation image
- translation arrow
- rotation (turn)
- clockwise
- counterclockwise
- turn centre
- reflection (flip)
- reflection image
- transformation
- tiling pattern
- tessellation
- tessellate
- ordered pair
- coordinates

- Describe the figures you see.
- In each picture, how could you move one figure to make it coincide with another figure?

LESSON 1

Translations

A firefighter slides down a pole.

A flag slides up a pole.

A child slides down a playground slide.

Which other ways do people or objects slide?

Explore

You will need Pattern Blocks, grid paper, and a ruler.

➤ Choose a Pattern Block.
Place it on the grid paper.
Trace the block.
Slide the block in a straight line.
Do not turn the block.
Use a ruler if it helps.
Trace the block in its new position.
Remove the block.
Have your partner write down how the block moved.

➤ Take turns to move a block and describe how it moved.

The straight line can be vertical, horizontal, or diagonal.

228 LESSON FOCUS | Draw and recognize translation images.

Show and Share

Trade your descriptions of how a block moved,
with a pair of classmates.
Follow your classmates' directions for moving a block.
Trace the block before and after you move it.
Compare drawings.
If they do not match, try to find out why.

Connect

When a figure moves along a straight line, without turning,
it is **translated** from one position to another.
The movement is a **translation** or a slide.
When we draw the figure in its new position,
we draw a **translation image** of the figure.
The translation is described by the numbers
of squares moved right or left and up or down.
This figure has been translated 5 squares right
and 4 squares down.

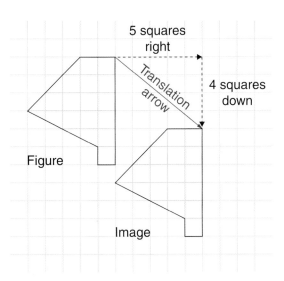

We say how many squares right or left before we say how many up or down.

The **translation arrow** shows how the figure moved.
The arrow joins matching points on the figure and its image.
A figure and its translation image are congruent.
They face the same way.

Practice

1. Does each picture show a translation?
   How do you know?

   a)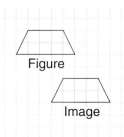

   b)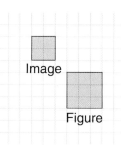

   c)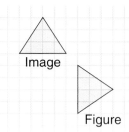

   d)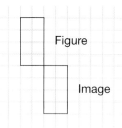

   If a picture does show a translation, describe it.

2. Copy each figure on grid paper.
   Translate the figure using the given translation.
   Draw the image and a translation arrow.

   a) 7 squares left and 3 squares up
   b) 5 squares right and 4 squares down

   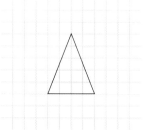
   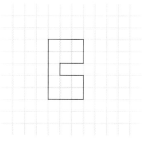

3. Write the translation that moved each figure to its image.

   a)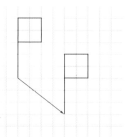
   b)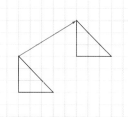
   c)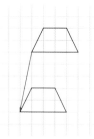

4. Draw this figure on grid paper.
   Translate it 3 squares left and 5 squares up.
   Draw the image.
   Label it Image A.
   Translate Image A 5 squares left and 3 squares up.
   Draw the new image.
   Label it Image B.
   Which translation would move the
   original figure directly to Image B? Explain.

5. Draw a figure on dot paper.
   Translate the figure in any direction.
   Draw its image.
   Continue to translate the figure or its image
   to make a pattern.
   Describe the translations you used.

6. Copy this figure and its translation image on grid paper.

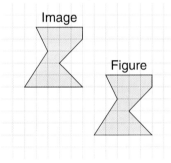

a) How many different translations would move
   the figure to its image?
b) What if you tried two translations, as in question 4?
   How many different pairs of translations
   would move the figure to its image?

Reflect

Use grid paper. Draw a figure and its
translation image. Explain how you know
your picture shows a translation.

Numbers Every Day

Calculator Skills

Find 3 consecutive
even numbers that
have a sum of 42.
Find the product
of the 3 numbers.

ASSESSMENT FOCUS | Question 5

Rotations

A bicycle wheel turns about the centre of the wheel.

What other examples are there of things that turn? Explain how they turn.

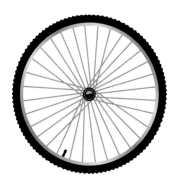

Explore

You will need several pieces of paper, tracing paper, a ruler, a compass, and scissors.

➤ Use a ruler.
   Draw a figure in the centre of a piece of paper.
➤ Use tracing paper.
   Draw a congruent figure on another piece of paper.
   Cut out this figure.
   Place it on top of the first figure you drew.
➤ Put your compass point at one vertex.
   Turn the figure to a new position.
   Draw the figure in its new position.
   Label this figure Image A.
➤ Return your figure to its original position.
   Turn the figure in the opposite direction.
   Draw the figure in its new position.
   Label this figure Image B.
➤ In each case, how does the original figure compare with its image?

232 LESSON FOCUS | Draw and recognize rotation images.

Show and Share

Compare your picture and ideas with another pair of classmates. Did you have the same ideas about how a figure compares with its image after a rotation? Explain.

Connect

When a figure turns about a point, it is **rotated** from one position to another.

The movement is a **rotation**, or turn. When we draw the figure in its new position, we draw a **rotation image** of the figure.

After 1 complete turn, a figure is back to where it started.

A figure can rotate **clockwise**:

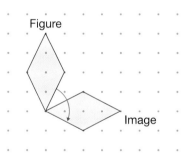

A figure can rotate **counterclockwise**:

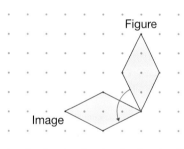

Any turn less than 1 complete turn is a fraction of a turn.

This figure has turned a $\frac{1}{4}$ turn clockwise, about vertex A. This point is called the **turn centre**.

This figure has turned a $\frac{1}{4}$ turn counterclockwise, about vertex B.

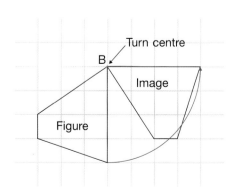

Unit 7 Lesson 2 **233**

A rotation is described by:
- the direction of the turn (clockwise or counterclockwise),
- the fraction of the turn, and
- the turn centre

A figure and its rotation image are congruent.
The figure and its image face different ways for any rotation that is less than 1 complete turn.

Practice

1. Which pictures show a rotation?
   How do you know? Describe the rotation.

   a)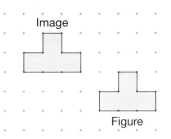

   b)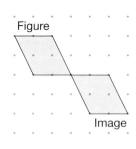

   c)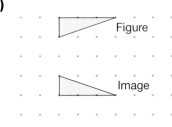

   d)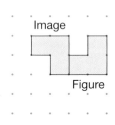

2. Did any of the pictures in question 1 show a translation?
   If so, identify the picture and describe the translation.

3. Copy this figure.
   Trace this figure on tracing paper.
   Use the tracing to rotate the figure.
   Rotate the figure:
   a) a $\frac{1}{4}$ turn clockwise about vertex A
   b) a $\frac{3}{4}$ turn counterclockwise about vertex A
   What do you notice about the rotation images?

   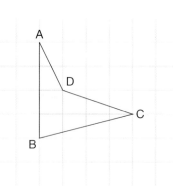

234 Unit 7 Lesson 2

4. Copy this figure.
   Use tracing paper to rotate the figure.
   Rotate the figure:
   a) a $\frac{1}{2}$ turn clockwise about vertex E
   b) a $\frac{1}{2}$ turn counterclockwise about vertex E
   What do you notice about the rotation images?

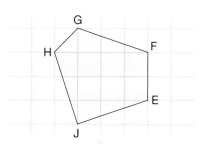

5. Copy this figure on grid paper.

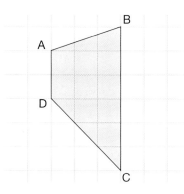

This figure is the image of a trapezoid that has been rotated a $\frac{1}{4}$ turn about one vertex.
Use tracing paper.
Draw all possible positions of the original trapezoid.
For each drawing, describe the rotation that produced the image.
How many different positions can you find?

6. Draw the capital letters of the alphabet.
   Use tracing paper.
   Rotate each letter about a point on the letter.
   Does any letter coincide with itself after less than 1 complete turn?
   If so, describe the rotation.

Reflect

When can two different rotations have the same image? Draw pictures to explain your answer.

Numbers Every Day

Number Strategies

Add the numbers in each set.
- 15, 25, 5, 10, 45
- 30, 10, 60, 20, 40, 20
- 3, 8, 9, 12, 6, 5, 6

LESSON 3

Reflections

A reflection can be used to make an interesting picture.
Is this person floating above the ground?
Where else do you see reflections?

Explore

You will need dot paper, a ruler, and a Mira.

➤ Draw a line through the centre of the dot paper.
  This is the mirror line.
➤ Use a ruler to draw a figure on one side of the line.
  Your partner draws the image of the figure in the mirror line.
  Use the Mira to check if the image is correct.
  If the image is not correct, keep the Mira in place and use it to draw the image.

LESSON FOCUS | Draw and recognize reflection images.

➤ Take turns to draw a figure and its image.
  Draw figures that touch the mirror line.
  Draw figures that cross the mirror line.
➤ In each case, how does the figure compare with its image?

Show and Share

Compare your pictures with those of another pair of classmates.
What is true about all figures and their images in a mirror line?

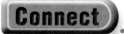

When a figure is **reflected** in a mirror, we see a **reflection image**.

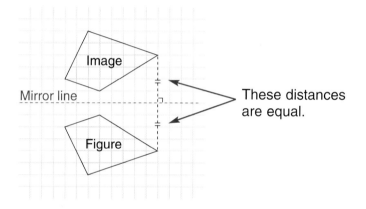

The line segment that joins a point to its image
is at right angles to the mirror line.
A point and its image are the same distance from the mirror line.
A mirror line can be vertical, horizontal, or at any
in-between position.
A figure and its reflection image are congruent.
They face opposite ways.
A reflection may be called a **flip**.
When a figure is reflected, it is flipped over.

A reflection, a rotation, and a translation are **transformations**.

Numbers Every Day

Number Strategies

Find 2 numbers with a difference of 9 and a product of 36.

Use a Mira when it helps.

1. Which pictures show a reflection?
   How do you know? Describe where the mirror line is.
   a)
   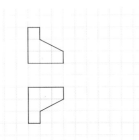
   b)
   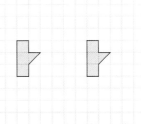
   c)
   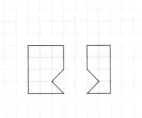
   d)
   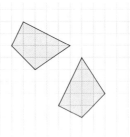

2. Copy each figure and mirror line on grid paper.
   Draw each reflection image.
   a)
   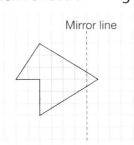
   b)
   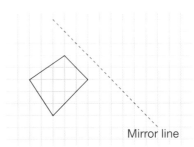

3. Each picture shows a figure and its reflection image.
   a)
   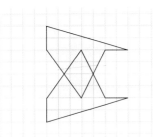
   b)
   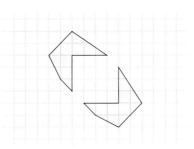

   Copy each picture.
   Draw the mirror line. Explain how you did this.
   How do you know the mirror line is correct?

4. Janine made a pattern using her initial and reflections.
   a) Describe how Janine made her pattern.
   b) Draw your initial on dot paper.
      Use reflections to make a pattern.
      Describe how you made your pattern.

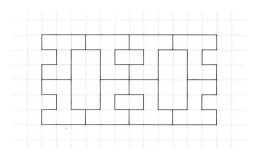

5. Use 1-cm grid paper.
   a) Draw a square.
      • Choose a translation.
        Translate the square. Label the image S.
      • Choose a mirror line.
        Reflect the square in the mirror line. Label the image R.
      • Choose a turn centre, a fraction of a turn, and the direction.
        Rotate the square about the turn centre. Label the image T.
      Compare the original square to each image.
   b) Repeat part a. This time, start with a rectangle.
      Compare the rectangle to each image.
   c) When you see a square and its image, can you always tell what the transformation was? Explain.
      Answer the same question for a rectangle and its image.

6. Describe the transformation that moves the figure to each image.
   Can you describe any movements in more than one way? Explain.
   a) Image A
   b) Image B
   c) Image C
   d) Image D

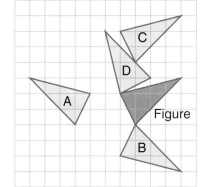

Reflect

When you see a figure and its image, how do you know if the picture shows a reflection, a rotation, or a translation?
Use diagrams to explain.

At Home

Look for an example of a transformation. Which transformation moves the figure to its image?

LESSON 4

Line Symmetry

A line of symmetry is a mirror line.
It divides a figure into 2 congruent parts.
How many lines of symmetry does each Pattern Block have?

Explore

You will need grid paper, Pattern Blocks, and a Mira.

➤ Fold the grid paper in half.
   Use Pattern Blocks.
   Make a design on one side
   of the fold line.
   Your design must touch
   the fold line.
   Trace around your design.
   Do not draw on the fold line.
➤ Open the paper. Use the Mira.
   Make a mirror image of
   the design on the other side
   of the fold line.
   Trace around the mirror image.
   Remove the Pattern Blocks.
➤ Find any lines of symmetry
   on the figure.

Show and Share

Show your work to another pair of students.
How did you find the lines of symmetry?
What did you notice about the fold lines?

LESSON FOCUS | Construct figures with line symmetry.

Connect

Here is one way to make a **symmetrical** figure.

> A symmetrical figure has one or more lines of symmetry.

➤ Fold a piece of paper.
  Draw a figure.
  Use the fold line as one side of the figure.
  Cut out the figure.

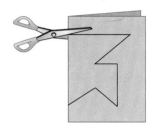

➤ Unfold the paper.
  The fold line is a line of symmetry.

Practice

1. Trace the figures that have line symmetry.
   Draw the lines of symmetry.

   a)    b)   c)

   d) 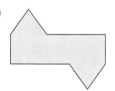   e)   f)

2. Choose a figure in question 1 that has line symmetry.
   You will cut out this figure from folded paper.
   Fold a piece of paper in half.
   Use scissors. Cut out a figure so that
   when the paper is unfolded, it matches
   the figure you chose in question 1.

3. One-half of a symmetrical figure is shown.
   The broken line is a line of symmetry.
   Copy the figure and the line of symmetry on dot paper.
   Complete the figure.
   a)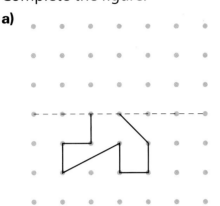
   b)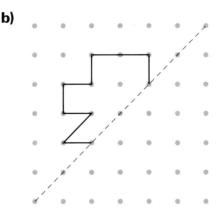

4. Use a geoboard and geobands.
   Divide the geoboard in half with a mirror line.
   Make a figure on one-half of the geoboard.
   Make its mirror image on the other half,
   so the two figures form a new figure.
   Is the new figure symmetrical? How do you know?

5. This figure does not have a line of symmetry.
   Copy the figure on grid paper.
   Add a square to it so it has a line of symmetry.
   How many different ways can you do this?
   Record each way on grid paper.

6. Fold a piece of paper in half.
   Cut the paper to make each figure
   when the paper is unfolded.
   a) An isosceles triangle
   b) A pentagon
   Tell about the figures you created.

Reflect

How can you construct a figure that has line symmetry?
Use pictures, numbers, or words to explain.

Numbers Every Day

Number Strategies

Order the numbers in each set from least to greatest.

• 5.37, 7.35, 3.75, 5.73, 7.53
• 0.11, 0.51, 0.15, 0.09, 1.01
• 4.88, 3.98, 1.78, 8.14, 9.38

LESSON 5

Exploring Tiling

The design on this fabric is an area pattern. There are no gaps between the figures. None of the figures overlap.

Choose one figure to be the start figure. Describe each other figure as an image of the start figure after a transformation.

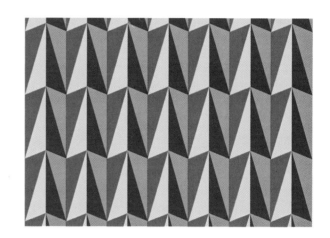

Explore

You will need Pattern Blocks, square dot paper, and triangular dot paper.

➤ Choose a Pattern Block.
  Try to cover a piece of paper
  with copies of the block,
  so there are no gaps.
  Copy the pattern on dot paper.
  Choose one block as the start figure.
  Explain how you could make
  the pattern using flips, slides, or turns.
➤ Repeat the activity using a
  different Pattern Block.

Show and Share

Which Pattern Blocks covered the paper with no gaps?
Which Pattern Blocks left gaps?
Trade patterns with another pair of classmates.
Use flips, slides, and turns to describe
your classmates' pattern.

LESSON FOCUS | Use polygons to tile and tessellate.

243

Connect

A **tiling pattern** covers a surface with figures.
There are no gaps or overlaps.

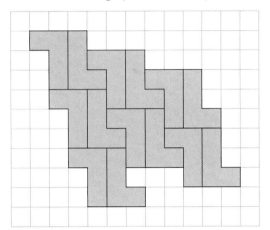

A tiling pattern with all figures congruent is a **tessellation**.

The hexagon was rotated a $\frac{1}{2}$ turn, about the dot shown. The figure formed by a hexagon and its rotation image was translated 3 squares right and 1 square down. To get the second row, the figure was translated 1 square right and 4 squares down.

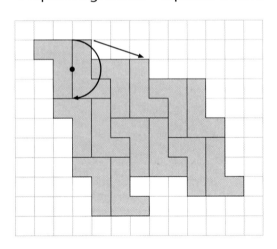

A figure **tessellates** when congruent copies of it cover a surface with no gaps or overlaps.

Practice

1. Use tracing paper.
   Trace several copies of each octagon.
   Then cut them out.
   Does a regular octagon tessellate?
   Does an irregular octagon tessellate?
   Explain.

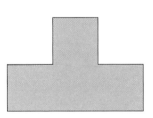

2. Copy this pentagon on grid paper.
   How many different tiling patterns
   can you make using this pentagon?
   Describe the transformations you could use
   to make each pattern.

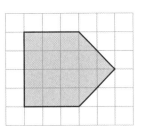

3. You will need grid paper and
   4 congruent squares.
   A tetromino is made with 4 congruent squares.
   Each square must align with at least one other square
   along one edge.

   This is a tetromino.   This is not a tetromino.

      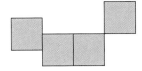

   a) How many different tetrominoes can you find?
      Record each tetromino on grid paper.
   b) Select one of your tetrominoes.
      Does it tessellate? Explain.

4. Use grid paper.
   Draw a figure.
   Transform the figure to create a tiling pattern.
   Describe the transformations you used.

5. Create a pentagon you think will tessellate.
   Trace several copies of the figure.
   Then cut them out.
   Does the figure tessellate?
   Tell why or why not.

Numbers Every Day

Number Strategies

Write each number as a decimal.

- thirty and fifty-three hundredths
- thirty-five and three-tenths
- twenty-two hundredths
- twenty and two-hundredths
- twenty-two and two-tenths

Reflect

How can you use transformations
to describe a tessellation?
Use pictures and words to explain.

LESSON 6

Strategies Toolkit

Explore

You will need Pattern Blocks and a Mira.

Choose 3 Pattern Blocks, 2 the same and 1 different.
Arrange the 3 blocks to make a figure with exactly
1 line of symmetry.
Each block must touch at least one other block.
Trace the figure.
Draw a dotted line to show the line of symmetry.

Show and Share

Describe the strategy you used to solve the problem.
Could you make more than one figure? Explain.

Connect

You will need pentominoes, 2-cm grid paper, and a Mira.
Choose 2 different pentominoes.
Arrange the pentominoes to create a figure
with exactly 1 line of symmetry.
Trace the figure and show the line of symmetry.

Strategies

- Make a table.
- Use a model.
- Draw a diagram.
- Solve a simpler problem.
- Work backward.
- Guess and check.
- Make an organized list.
- Use a pattern.
- Draw a graph.

What do you know?
- Use 2 different pentominoes.
- Arrange the pentominoes to make a figure.
- The figure must have exactly 1 line of symmetry.

Think of a strategy to help you solve this problem.
- You can use **guess and check** to find a figure with exactly 1 line of symmetry.

246 LESSON FOCUS | Interpret a problem and select an appropriate strategy.

Arrange the pentominoes to make a figure.
Use a Mira to check for lines of symmetry.
If the figure has no lines of symmetry
or more than one line of symmetry,
try a different arrangement to make a new figure.

Check your work.
Does your figure have exactly 1 line of symmetry?
How do you know?

Practice

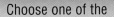

Choose one of the **Strategies**

1. Draw mirror lines to divide a piece of grid paper in 4 congruent sections.
   Draw Figure A in one section.
   Reflect Figure A in one of the mirror lines.
   Label the image B.
   Reflect Image B in the other mirror line.
   Label the image C.
   Describe a transformation that would move Figure A directly onto Image C.
   How many different transformations can you find?

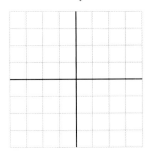

2. Repeat question 1.
   This time divide the paper in 3 sections.

Reflect

How does guess and check help you solve a problem?
Use pictures and words to explain.

Unit 7 Lesson 6   **247**

LESSON

Coordinate Grids

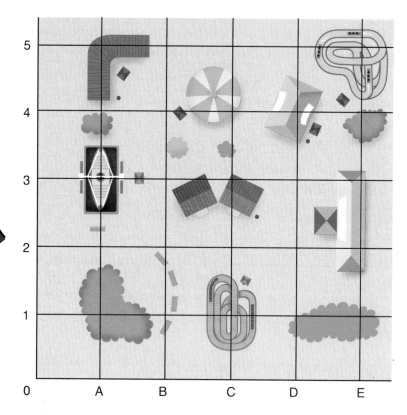

Here is a map of an amusement park.
The roller coaster is at C1.
What are the coordinates of the water ride?
The swinging ship?

You will need 1-cm grid paper and a ruler.

➤ Draw and label a grid like this:
➤ Draw a figure on your grid.
  Place each vertex where two grid lines meet.
➤ With your partner, find a way to describe your figure so that someone else can draw it without seeing it.
➤ Write down your description.

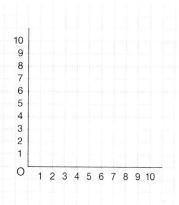

Show and Share

Talk with another pair of classmates.
Trade ideas for describing the position of a figure on a grid.
Trade descriptions with your classmates.
Use your classmates' description to draw their figure.

248 LESSON FOCUS | Plot ordered pairs on a grid and identify points.

➤ To describe the position of a point on a grid, we use an **ordered pair**.
It describes the translation that moves the point from O on the grid.

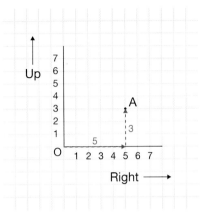

The first coordinate tells how far you move right. The second coordinate tells how far you move up.

From O, to reach point A, we move 5 units right and 3 units up.
We write these numbers in brackets: (5,3)
The ordered pair (5,3) tells how far to move right and up to get from O to A.

We say: A has coordinates (5,3).
We write: A(5,3)

The point O has coordinates (0,0) because you do not move anywhere to plot a point at O.

➤ When the numbers in an ordered pair are large, we use a scale on the grid.
On this grid, 1 square represents 10 units.

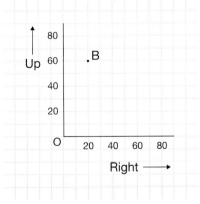

"Coordinates" is another name for "ordered pair."

To plot point B(20,60):
Start at O.
Move 2 squares right.
Move 6 squares up.

Practice

1. Match each ordered pair with a letter on the grid.
   a) (1,5)
   b) (5,1)
   c) (0,7)
   d) (7,0)

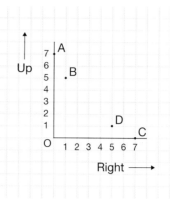

2. Draw and label a grid. Plot each point on the grid.
   a) P(2,7)   b) Q(6,5)   c) R(1,4)   d) S(0,3)

3. Mr. Kelp's class went to the Vancouver Aquarium. Angel drew this map of the aquarium site. Write the ordered pair for each place.
   a) Amazon Jungle Area: A
   b) Beluga whales: B
   c) Carmen the Reptile: C
   d) Entrance: E
   e) Frogs: F
   f) Sea otters: S
   g) Sharks: H

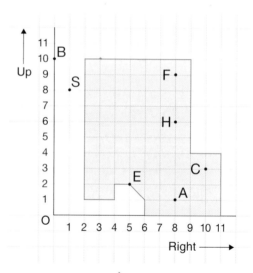

4. Use the map in question 3.
   a) To get to the Pacific Canada Pavilion at point P:
   You move 1 square left and 3 squares up from the entrance, E.
   What are the coordinates of P?
   b) To get to the Clam Shell Gift Shop at point G:
   You move 5 squares left and 4 squares down from the sharks, H.
   What are the coordinates of G?

5. Copy this grid.
   a) Plot each point on the grid.
   A(10,5)   B(5,15)   C(10,25)
   D(20,25) E(25,15) F(20,5)
   b) Join the points in order. Then join F to A.
   What figure have you drawn?
   Is it a regular figure? How do you know?
   c) What are the attributes of this figure?

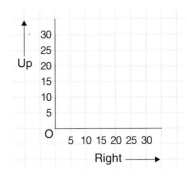

6. Draw and label a grid.
   a) Plot the points A(5,1) and B(5,5).
   Join the points.
   b) Find point C so that △ABC is isosceles.
   How many different ways can you do this?
   Draw each way you find.
   Write the coordinates of C.
   How do you know each triangle is isosceles?
   c) Find point D so that △ABD is scalene.
   Show 3 different scalene triangles.
   Write the coordinates of D.
   How do you know each triangle is scalene?
   d) Can you find point E so that △ABE is equilateral?
   If you can, draw △ABE.
   Write the coordinates of E.
   If you cannot, explain why △ABE cannot be drawn.

7. Draw and label a grid.
   a) Draw a square on the grid so that each vertex is at a point where grid lines meet.
   What are the coordinates of the vertices?
   b) Draw the smallest square you can.
   What are the coordinates of its vertices?
   c) Draw the largest square you can.
   What are the coordinates of its vertices?

Numbers Every Day

Number Strategies

Will each product be greater than or less than 812 × 19? How do you know?

850 × 10
800 × 20
850 × 15
810 × 20

Reflect

What is an ordered pair?
Use diagrams in your explanation.

ASSESSMENT FOCUS | Question 6

Unit 7 Show What You Know

LESSON

1 2 3

1. Describe a transformation that would move Figure A to each image.
   a) Image B
   b) Image C
   c) Image D
   d) Image E

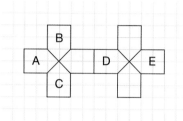

1 2 3 7

2. a) Describe the translation that moves:
   i) Figure B to Image A
   ii) Figure D to Image C

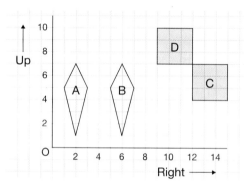

   b) Which other transformation would move each figure above to its image?

4

3. One-half of a symmetrical figure is shown. The broken line is a line of symmetry. Copy the figure and line of symmetry on dot paper. Complete the figure.

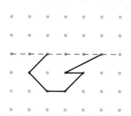

1 3 7

4. a) Write the ordered pairs for the vertices A, B, C, and D.
   b) Vertex B is translated 2 units right and 5 units up. What are the coordinates of the translation image of B?
   c) Vertex D is reflected in a mirror line through AB. What are the coordinates of the reflection image of D?
   d) Vertex A is rotated a $\frac{1}{4}$ turn clockwise about point B. What are the coordinates of the rotation image of A?

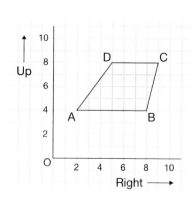

LESSON

5. You will need triangular grid paper and 6 green Pattern Blocks.
   Arrange all 6 triangles to make a figure.
   Where triangles meet, their sides must align.
   a) How many different figures can you make?
      Record each figure on grid paper.
   b) Select one of your figures.
      Does it tessellate? Explain.

6. Choose one figure.
   Explain how each other figure is the image after a transformation.

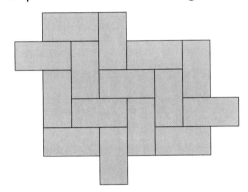

7. Plot each set of points on the same grid.
   Join the points as you plot them.
   a) (5,4), (4,5), (2,3), (1,3), (1,1), (4,1), (4,3), (2,3)
   b) (4,2), (5,2), (5,1), (8,1), (8,2), (11,5), (12,4)
   c) (5,2), (5,3), (8,3), (8,2)
   What did you draw?

8. Use question 7 as a guide.
   Draw a picture on a grid.
   List the coordinates of each vertex in order.
   Trade coordinates with a classmate.
   Draw your classmate's picture.

UNIT 7 Learning Goals

- ✓ recognize a slide (translation), a turn (rotation), and a flip (reflection)
- ✓ construct figures with one line of symmetry
- ✓ explore tiling patterns and tessellations
- ✓ plot and identify points on a coordinate grid

Unit 7   253

Unit Problem

Geometry in Art

You will need:
- Bristol board
- coloured paper
- a 10-cm square cut from 1-cm grid paper
- scissors

Part 1

Make a pattern for a tile that tessellates.
- Draw a figure on the grid paper square.
  One side of the figure must coincide with one side of the square.

- Cut out the figure.
- Slide the figure across the square and tape it to the opposite side. The edges must line up with no gaps or overlaps.
- Continue to cut figures and tape them to the opposite side until your pattern is the way you want it.

Part 2

Make a tessellation.
- Use your pattern to cut tiles from coloured paper.
- Glue the tiles onto Bristol board.

Part 3

Write about your tessellation.
- Does your tessellation have line symmetry?
  How do you know?
- How could you use transformations to make the tessellation?
  Describe as many ways as you can find.

Check List

Your work should show
- ✓ a pattern for a tile that tessellates
- ✓ a tessellation of coloured tiles on Bristol board
- ✓ a description of whether your tessellation has line symmetry
- ✓ an explanation of how you could use transformations to make the tessellation

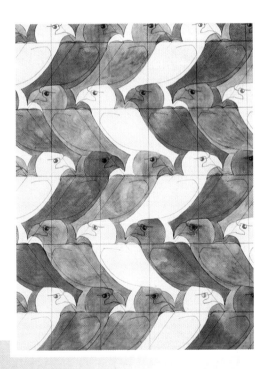

Reflect on the Unit

What do you know about transformations and tessellations?
Use pictures and words to explain.

Cross Strand Investigation

Rep-Tiles

You will need Pattern Blocks.

Part 1

A **rep-tile** is a polygon that can be copied and arranged to form a larger, similar polygon.

These are rep-tiles: These are not rep-tiles:

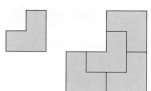

➤ Which Pattern Blocks are rep-tiles? How did you find out?

Part 2

Choose a block that is a rep-tile.
Do not use orange or green blocks.
Build a growing pattern.
Record the pattern.

➤ Choose one Pattern Block that is a rep-tile. This is Frame 1.
➤ Now take several of the same type of block. Arrange the blocks to form a similar polygon. This is Frame 2.
Continue to arrange blocks to make larger similar polygons.
The next largest polygon is Frame 3.

Similar figures have the same shape. They may not have the same size.

➤ Suppose the side length of the green Pattern Block is 1 unit.
  Find the perimeter of each figure.
➤ Suppose the area of the green Pattern Block is 1 square unit.
  Find the area of each frame.
  Copy and complete the table.

Frame	Number of Blocks	Perimeter	Area
1	1		
2			

Part 3

➤ What patterns can you find in the table?
➤ How many blocks would you need to build Frame 7?
  How do you know?
➤ Predict the area and the perimeter of Frame 9.
  How did you make your prediction?

Display Your Work

Record your work.
Describe the patterns you found.

Take It Further

Draw a large polygon you think is a rep-tile.
Trace several copies.
Cut them out.
Try to arrange the copies to make a larger similar polygon.
If your polygon is a rep-tile, explain why it works.
If it is not, describe how you could change it to make it work.

257

UNIT 8

Fractions and In the Garden

Brian and Samantha are planning a garden. What fraction of the garden will they plant with flowers? Vegetables?

Learning Goals

- model, compare, and order fractions, improper fractions, and mixed numbers
- explore equivalent fractions and decimals
- explore patterns involving fractions
- relate fractions to division and to decimals
- estimate decimal products and quotients
- multiply decimals with tenths and with hundredths
- divide decimals with tenths
- pose and solve problems involving decimals and fractions

Flower seeds
10 packs for $1.50
Zucchini seeds
$1.09 per pack
Tomato plants
$0.85 each
Pumpkin seeds
$0.99 per package

Decimals

Key Words

equivalent fractions

proper fraction

mixed number

improper fraction

quotient

divisor

dividend

Brian and Samantha will buy seeds and plants.
- What is the cost of 10 packages of zucchini? 100 tomato plants?
- About how much will 1 package of flower seeds cost? How could you find the exact amount?

Equivalent Fractions

$\frac{6}{12}$ of the stickers have gone.

$\frac{1}{2}$ of the stickers are left.

$\frac{1}{2}$ and $\frac{6}{12}$ name the same amount.

They are **equivalent fractions**.
What does each numerator represent?
What does each denominator represent?

Explore

You will need Colour Tiles or congruent squares.
Your teacher will give you a copy of this rectangle.

➤ Place the tiles on the rectangle so that:
  - $\frac{1}{6}$ of the rectangle is red.
  - $\frac{2}{3}$ of the rectangle is blue.
  - $\frac{1}{9}$ of the rectangle is green.
  - The rest of the rectangle is yellow.

  Record your work on the rectangle.

➤ How many ways can you describe the fraction of the rectangle that is red? Blue? Green? Yellow? Record each way.

260   LESSON FOCUS | Find equivalent fractions and investigate patterns among them.

Show *and* Share

Share your work with another pair of students.
Compare the fractions you wrote for each colour.
How did you know which fractions to write?
Describe any patterns you see in the fractions for each colour.

Connect

This rectangle was made with Colour Tiles.

What fraction of the rectangle is green?
How many different fractions can you write?

➤ There are 24 tiles in the rectangle.
  12 tiles are green.
  $\frac{12}{24}$ of the rectangle is green.

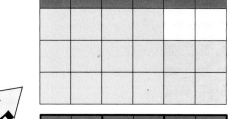

➤ There are 12 groups of 2 tiles.
  6 groups are green.
  $\frac{6}{12}$ of the rectangle is green.

➤ There are 8 groups of 3 tiles.
  4 groups are green.
  $\frac{4}{8}$ of the rectangle is green.

➤ There are 6 groups of 4 tiles.
  3 groups are green.
  $\frac{3}{6}$ of the rectangle is green.

 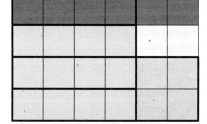

➤ There are 4 groups of 6 tiles.
  2 groups are green.
  $\frac{2}{4}$ of the rectangle is green.

➤ Finally, there are 2 groups of 12 tiles.
1 group is green.
$\frac{1}{2}$ of the rectangle is green.

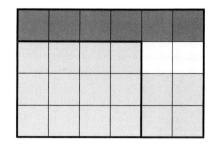

$\frac{1}{2}, \frac{2}{4}, \frac{3}{6}, \frac{4}{8}, \frac{6}{12}$, and $\frac{12}{24}$ represent the same amount.
They are equivalent fractions.

➤ There are patterns in the equivalent fractions.

$\frac{1}{2}, \frac{2}{4}, \frac{3}{6}, \frac{4}{8}, \frac{6}{12}, \frac{12}{24}$

The numerators are multiples of the least numerator, 1.

The denominators are multiples of the least denominator, 2.

There are other fractions equivalent to $\frac{1}{2}$.
I could use different rectangles to find equivalent fractions with denominators such as 10, 14, and 16.

Practice

Use Colour Tiles or grid paper when they help.

1. What fraction of each figure is blue?
   How many different fractions can you write each time?

   a)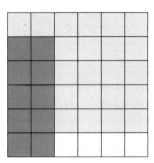

   b)

2. Find as many equivalent fractions as you can for each picture.

a)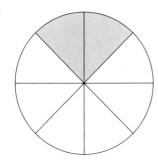
b)
c)

3. Write 3 fractions that are equivalent to $\frac{2}{5}$. Explain how you did it.

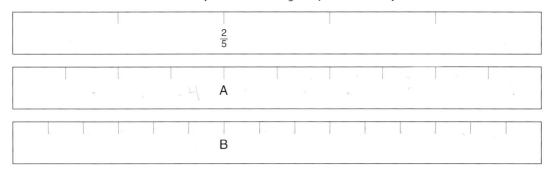

4. Draw a picture to show each pair of equivalent fractions.

a) $\frac{1}{4}, \frac{3}{12}$
b) $\frac{2}{3}, \frac{8}{12}$
c) $\frac{3}{5}, \frac{12}{20}$

5. Rhonda, Apak, Kayla, and Sunil each ordered a large pizza.
Each pizza was cut into slices of equal size.
Rhonda's pizza had 6 slices, Apak's had 8,
Kayla's had 10, and Sunil's had 12.
Each student ate her or his own pizza.
Rhonda ate 3 slices, Apak ate 4, Kayla ate 5, and Sunil ate 6.
Sunil says he ate the most because he ate 6 pieces.
Rhonda says everyone ate the same amount.
Who is correct? Use pictures, numbers,
or words to explain.

Reflect

How can you tell if two fractions are equivalent?
Use numbers, pictures, or words to explain.

Numbers Every Day

Calculator Skills

Use 2 ways to find the quotient of 156 ÷ 6 without pressing the ÷ key.

L E S S O N 2

Fractions and Mixed Numbers

$\frac{4}{8}$ is a **proper fraction**.

$1\frac{1}{2}$ is a **mixed number**.

$\frac{12}{8}$ is an **improper fraction**.

How are these numbers related?

Explore

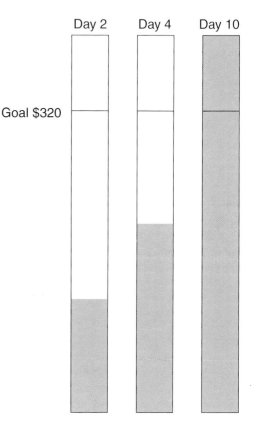

You will need Cuisenaire rods or a ruler.

Samuel's class was raising money for charity. Its goal was to raise $320.

The pictures show the total amount of money raised after 2 days, 4 days, and 10 days.

What fraction of the goal did the class reach on Day 2? On Day 4? On Day 10?

Show and Share

Share your work with another pair of students. What strategies did you use to find the fractions? How many ways can you describe the fraction of the goal reached on Day 10?

Connect

Suppose  = 1 whole.

What fraction does a blue rod represent?

Two pink rods fit on the blue rod, with some of the blue rod left over.
There are 2 wholes with some left over.
The length left over is the length of a tan rod.
One tan rod is $\frac{1}{4}$ of one pink rod.

1 blue rod represents $\frac{9}{4}$.

1 blue rod represents $2\frac{1}{4}$.

$2\frac{1}{4}$ and $\frac{9}{4}$ are equivalent.

Practice

Use Cuisenaire rods or paper strips when they help.

1. Which scoop would you use to measure each amount?
   How many scoops would you need?

   a) $2\frac{1}{4}$ cups b) $2\frac{1}{2}$ cups
   c) $1\frac{2}{3}$ cups d) $3\frac{1}{2}$ cups

2. Look at question 1. How many different ways can you use the scoops to measure each amount? Record each way.

Number Strategies

Find each product.

$12 \times 16 = \square$
$24 \times 16 = \square$
$12 \times 32 = \square$
$24 \times 32 = \square$

3. What fraction does each picture represent?
   Write an equivalent fraction for each.
   a)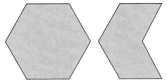
   b)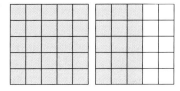
   c)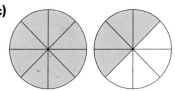

4. Are the fractions in each pair equivalent?
   How do you know?
   a) $3\frac{2}{3}$ and $\frac{11}{3}$
   b) $\frac{13}{8}$ and $1\frac{5}{8}$
   c) $2\frac{3}{4}$ and $\frac{9}{4}$

5. Kendra mowed her lawn in $2\frac{1}{2}$ h.
   Mario mowed his lawn for $\frac{1}{4}$ h, then stopped.
   He did this 7 times.
   Who spent the most time mowing the lawn?
   How do you know?

6. Carlo baked pies for a party.
   He cut some pies into 6 pieces
   and some into 8 pieces.
   After the party, more than $2\frac{1}{2}$
   but less than 3 pies were left.
   How much pie might have
   been left?
   Show how you know.

7. How can you find out if $2\frac{1}{2}$ and $\frac{10}{4}$
   name the same amount?
   Use words, numbers, and pictures to explain.

Reflect

When you see an improper fraction,
how can you write it as a mixed number?

At Home

When do you use fractions
and mixed numbers at home?
Explain.

LESSON 3

Comparing and Ordering Fractions

Explore

Your teacher will give you a large copy of these number lines.

The first number line shows halves. The second shows thirds, and so on.

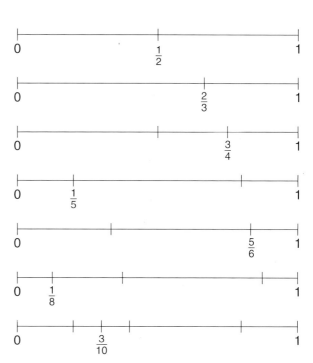

➤ The number lines are incomplete. Mark and label the lines with the missing fractions. Use any materials that help.
➤ Describe any patterns in the completed lines.
➤ Which fraction in each pair is greater? How do you know?
  • $\frac{2}{3}$ or $\frac{2}{5}$
  • $\frac{3}{8}$ or $\frac{4}{10}$

Show and Share

Show your work to another pair of students. How did you know where to place the fractions on the number lines? Share the patterns you found. How did you decide which fraction was greater?

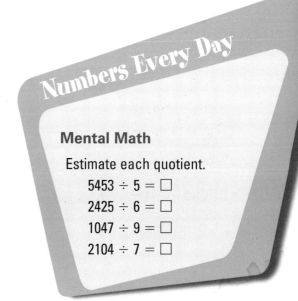

Numbers Every Day

Mental Math

Estimate each quotient.
5453 ÷ 5 = ☐
2425 ÷ 6 = ☐
1047 ÷ 9 = ☐
2104 ÷ 7 = ☐

LESSON FOCUS | Compare and order fractions.

Connect

Here are 3 ways to order fractions.

➤ Order $\frac{3}{4}$, $\frac{3}{5}$, and $\frac{5}{8}$ from least to greatest.

Fold or measure, then colour, 10-cm strips of paper.

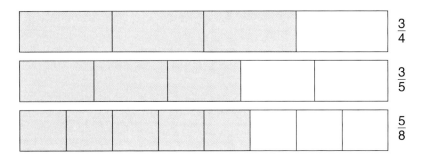

The least fraction is the shortest coloured strip.
The order from least to greatest is $\frac{3}{5}$, $\frac{5}{8}$, and $\frac{3}{4}$.

➤ Order $\frac{2}{5}$, $\frac{1}{4}$, and $\frac{5}{8}$ from least to greatest.

Draw number lines.

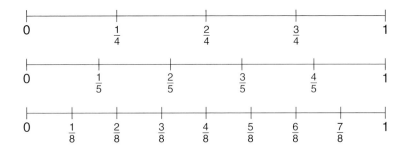

$\frac{1}{4}$ is to the left of $\frac{2}{5}$.

$\frac{2}{5}$ is to the left of $\frac{5}{8}$.

The order from least to greatest is $\frac{1}{4}$, $\frac{2}{5}$, and $\frac{5}{8}$.

➤ Compare pairs of fractions with the same numerator or denominator.

Compare $\frac{3}{8}$ and $\frac{3}{5}$.
Both fractions have the same number of parts, 3.
Since one-fifth is greater than one-eighth, $\frac{3}{5}$ is greater than $\frac{3}{8}$.

Compare $\frac{3}{4}$ and $\frac{5}{8}$.
Find equivalent fractions with eighths.
$\frac{3}{4} = \frac{6}{8}$
$\frac{6}{8} > \frac{5}{8}$, so $\frac{3}{4} > \frac{5}{8}$

Practice

1. Use three 12-cm strips of paper.
   Show thirds on one strip.
   Show halves on another strip.
   Show quarters on the third strip.
   Use the strips to order these fractions from least to greatest.
   $\frac{1}{3}, \frac{1}{2}, \frac{1}{4}$

2. Use three 10-cm strips of paper.
   Show halves on one strip.
   Show tenths on another strip.
   Show fifths on the third strip.
   Use the strips to order these fractions from least to greatest.
   $\frac{1}{2}, \frac{7}{10}, \frac{4}{5}$

3. Use three 12-cm strips of paper.
   Mark each strip with appropriate fractions.
   Use the strips to order these fractions from least to greatest.
   $\frac{5}{6}, \frac{2}{3}, \frac{7}{12}$

4. Use 3 copies of a 12-cm number line like the one below.

   ⊢————————————————⊣
   0   1

   Show eighths on one line.
   Show sixths on another line.
   Show quarters on the third line.
   Use the number lines to order these fractions from least to greatest.
   $\frac{7}{8}, \frac{4}{6}, \frac{3}{4}$

Unit 8 Lesson 3 **269**

5. Use 3 copies of a 12-cm number line like the one below.

   |————————————————————————|
   0   1

   Show halves on one line.
   Show thirds on another line.
   Show sixths on the third line.
   Use the number lines to order these fractions
   from least to greatest.

   a) $\frac{2}{2}, \frac{1}{3}, \frac{3}{6}$ b) $\frac{1}{6}, \frac{2}{3}, \frac{5}{6}$

6. Use grid paper.
   Draw pictures to represent 3 fractions that are greater than $\frac{3}{5}$.
   Each fraction should have a different denominator.

7. A quilt has 20 patches.
   One-quarter of the patches are yellow,
   $\frac{3}{5}$ are green, and the rest are red.
   What colour are the greatest number of patches?
   The least number of patches? Show how you know.

8. Jessica and Ramon each have the same length of ribbon.
   Jessica cut her ribbon into eighths.
   Ramon cut his ribbon into twelfths.
   Jessica sold 6 pieces and Ramon sold 8.
   Who sold the most ribbon? How did you find out?

9. Which is greater, $\frac{2}{3}$ or $\frac{2}{5}$? How do you know?

10. Compare the fractions in each pair.
    Copy each statement. Write >, <, or =.
    How did you decide which symbol to choose?

    a) $\frac{4}{5} \square \frac{4}{10}$ b) $\frac{3}{8} \square \frac{2}{8}$ c) $\frac{2}{3} \square \frac{4}{6}$ d) $\frac{1}{4} \square \frac{1}{3}$

Reflect

Use pictures, words, or numbers to explain which is greater,
$\frac{7}{8}$ or $\frac{3}{4}$.

Order Up!

Your teacher will give you a set of fraction cards and a set of game cards.

- ➤ Shuffle the fraction cards.
  Place them face down in a pile.
- ➤ Take turns to draw a card.
  Record the fraction on your game card.
  The goal is to write the fractions on your game card in order from least to greatest.
  Once you have recorded a fraction on your game card, you cannot erase it.
  Return the card to the bottom of the pile.
- ➤ If you cannot record the fraction on a card, return the card to the bottom of the pile.
  The next player takes a turn.
- ➤ The first player to fill her or his game card correctly wins.

LESSON 4

Relating Fractions to Decimals

Many vegetable gardens have similar vegetables planted in the same area.

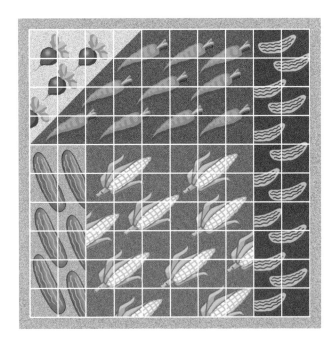

Explore

You will need Base Ten Blocks and grid paper.

Use rods and unit cubes to design a vegetable garden.
Use a flat to represent the whole garden.
Each vegetable is to occupy a separate region of the garden.
The garden must have:
- more carrots than corn
- more onions than zucchini
- all of the land planted with one of these vegetables

Record your vegetable garden design on grid paper.

➤ Write the fraction of your garden planted with each vegetable in as many ways as you can.

➤ How many ways can you use a decimal to describe the fraction of the garden that is planted with each kind of vegetable? Record each way.

272   LESSON FOCUS | Relate fractions with denominators of 10 and 100 to decimals.

Show and Share

Share your results with another pair of students.
How did you find the fractions and decimals?
Which fractions and decimals name the same amount?
How do you know?

Connect

This is Jake and Willa's design of a flower garden.

$\frac{1}{4}$, or $\frac{25}{100}$, of the garden is planted with roses.

$\frac{1}{4}$, or $\frac{25}{100}$, of the garden is planted with tulips.

$\frac{3}{10}$, or $\frac{30}{100}$, of the garden is planted with lilies.

$\frac{2}{10}$, or $\frac{20}{100}$, of the garden is planted with daisies.

➤ You can write fractions with denominators of 10 and 100 as decimals.

$\frac{3}{10}$ is 3 tenths, or 0.3.

$\frac{15}{100}$ is 15 hundredths, or 0.15.

$\frac{125}{100}$ is 125 hundredths, or 1.25.

➤ If a fraction does not have a denominator of 10 or 100, try to find an equivalent fraction that does.

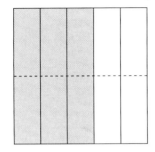

$\frac{3}{5}$ is equivalent to $\frac{6}{10}$.

$\frac{6}{10}$ is 6 tenths, or 0.6.

$\frac{3}{5}$ and 0.6 are equivalent.

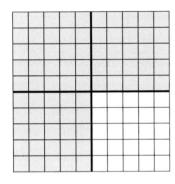

$\frac{3}{4}$ is equivalent to $\frac{75}{100}$.

$\frac{75}{100}$ is 75 hundredths, or 0.75.

$\frac{3}{4}$ and 0.75 are equivalent.

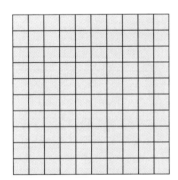 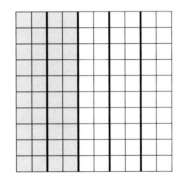

$1\frac{2}{5}$ is equivalent to $1\frac{4}{10}$.

$1\frac{4}{10}$ is one and 4 tenths, or 1.4.

$1\frac{2}{5}$ and 1.4 are equivalent.

Practice

1. Write a fraction and a decimal to describe:
   a) the shaded part of the grid
   b) the white part of the grid

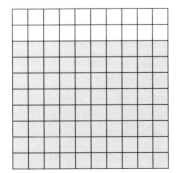

2. Vijay has $\frac{1}{20}$ of a dollar in his pocket. What coins might he have?

3. Use Base Ten Blocks to represent each fraction. Then write each fraction as a decimal.
   a) $\frac{1}{2}$ b) $\frac{3}{4}$ c) $\frac{9}{10}$
   d) $\frac{47}{100}$ e) $\frac{3}{5}$ f) $\frac{7}{5}$

4. Represent each fraction on a hundredths grid. Then write each fraction as a decimal.
   a) $\frac{1}{4}$ b) $\frac{4}{5}$ c) $\frac{3}{100}$

5. Draw a picture to represent each number. Then write each number as a decimal.
   a) $7\frac{1}{2}$ b) $2\frac{3}{4}$ c) $\frac{4}{5}$ d) $2\frac{7}{20}$

6. Copy and complete. Replace each □ with <, >, or = to make the statement true.
   a) $\frac{75}{100}$ □ $\frac{3}{4}$ b) 0.08 □ $\frac{8}{10}$ c) 0.6 □ $\frac{60}{100}$ d) $\frac{8}{5}$ □ $1\frac{3}{10}$

7. Write 2 equivalent fractions for each decimal.
   a) 0.5   b) 0.75   c) 0.40   d) 0.9

8. Two fractions were written as decimals, then added. The sum is 0.45.
   a) What might the decimals be?
   b) What might the fractions be? Show how you know.

9. Do $2\frac{3}{5}$ and 2.35 name the same amount? Use pictures and words to explain how you know.

Reflect

Which fractions can you write easily as a decimal? Use examples in your explanation.

Numbers Every Day

Number Strategies

Which of these show 62.56?
- six thousand two hundred fifty-six
- 62.0 + 0.56
- 6256 ÷ 10
- 625.6 ÷ 10
- 62.56 × 0
- 60 + 2.56

ASSESSMENT FOCUS | Question 8

LESSON 5

Fraction and Decimal Benchmarks

You have used fraction benchmarks to estimate with fractions.

| 0 | $\frac{1}{2}$ | 1 |

Is $\frac{3}{10}$ closer to 0, $\frac{1}{2}$, or 1? How do you know?

In this lesson, you will explore decimal benchmarks.

Explore

Jamal surveyed 100 people.

He asked this question:
What kind of animal would you most like to have as a pet?
The bar graph shows his results.

➤ What fraction of the people surveyed would like to have each animal as a pet?
➤ Write each fraction as a decimal.

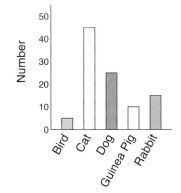

Most Popular Animal to Have as a Pet

Show and Share

Share your work with another pair of students.
How did you estimate each fraction? Each decimal?

276 LESSON FOCUS | Explore benchmarks for fractions and decimals.

Connect

You can use benchmarks to compare fractions and decimals by estimating.

Extend the fraction benchmarks to include quarters.
We can rename the benchmarks as decimals.

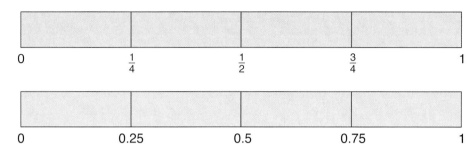

➤ Which benchmark is $\frac{5}{12}$ closest to?

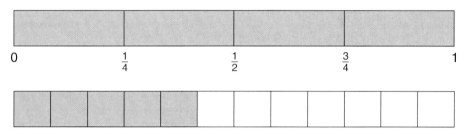

$\frac{5}{12}$ is greater than $\frac{1}{4}$ and less than $\frac{1}{2}$.
$\frac{5}{12}$ is closer to $\frac{1}{2}$.

➤ Which benchmark is 0.6 closest to?

0.6 is greater than 0.5 and less than 0.75.
0.6 is closer to 0.5.

Practice

1. Which benchmark is each fraction closest to? How do you know?

a)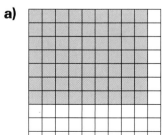

b)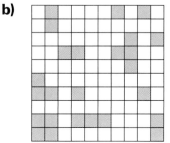

c)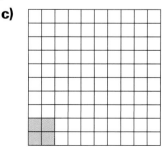

Unit 8 Lesson 5

2. Draw a picture for each decimal.
   What is the closest benchmark for each decimal?
   a) 0.3   b) 0.10   c) 0.75   d) 0.05

3. Write each fraction as a decimal.
   a) $\frac{2}{4}$ b) $\frac{9}{10}$ c) $\frac{37}{100}$ d) $\frac{3}{5}$
   Which decimal benchmark is each fraction closest to?

4. Use a copy of this number line.

   a) Place $\frac{7}{10}$ and $\frac{4}{5}$ on the number line.
      Which benchmarks are closest to $\frac{7}{10}$ and $\frac{4}{5}$?
   b) Can you place 5 decimals between
      the fractions on the number line?
      If so, mark and label each decimal. If not, explain why not.
   c) Use a hundredths grid to write 5 decimals between $\frac{7}{10}$ and $\frac{4}{5}$.

5. Use benchmarks to compare $\frac{4}{5}$ and 0.9.

6. Copy and complete the table.

Decimal	Lower Benchmark	Upper Benchmark	Nearest Benchmark
0.21			
0.09			
0.37			
0.63			
0.80			
0.99			

Reflect

Describe how using benchmarks can help you to compare and order fractions and decimals.

Numbers Every Day

Number Strategies

Find each difference.

17.5 − 8.3
0.85 − 0.07
21.2 − 10.3
10.01 − 6.51

LESSON 6
Relating Fractions to Division

Explore

Javier has 11 apples to share with a friend. How many apples will each person get?

Try to do this 2 different ways. How will you record your answer?

Show and Share

Share your answer with another pair of classmates. Did you get the same answer? How do you know?

Connect

Helena has 8 doughnuts to share among 5 people. How much will each person get?

Here are 2 ways to solve the problem.

➤ Use pictures.

Each person has 1 doughnut. There are 3 left over.
Divide each leftover doughnut in fifths.

There are $\frac{15}{5}$.
Each person gets $\frac{3}{5}$ of the leftover doughnuts.

Each person gets $1\frac{3}{5}$ doughnuts.

LESSON FOCUS | Relate fractions to division.

279

➤ Divide.

Eight doughnuts shared among 5 people is written as 8 ÷ 5, or $\frac{8}{5}$.

$\frac{8}{5}$ is an improper fraction.

The 3 left over are shared among 5 people.
This is written as 3 ÷ 5, or $\frac{3}{5}$.

We write 1 R3 as $1\frac{3}{5}$.

Any division statement can be written as a fraction.

$8 \div 5 = \frac{8}{5}$

If the quotient is greater than 1, it can be written as a mixed number.

$8 \div 5 = \frac{8}{5} = 1\frac{3}{5}$

Practice

1. Write each division statement as a fraction.
   Use grid paper. Draw a picture for each fraction.
   a) 2 ÷ 4   b) 3 ÷ 8   c) 4 ÷ 10   d) 5 ÷ 12

2. Write each division statement as a fraction.
   Use grid paper. Draw a picture for each fraction.
   a) 5 ÷ 4   b) 11 ÷ 8   c) 14 ÷ 10   d) 15 ÷ 12

3. Write each division statement as an improper fraction
   and as a mixed number.
   Draw a picture to show each fraction.
   a) 35 ÷ 6   b) 66 ÷ 5   c) 29 ÷ 3   d) 121 ÷ 10

4. Write each fraction as a division statement.
   a) $\frac{2}{3}$ b) $\frac{14}{9}$ c) $\frac{1}{8}$ d) $\frac{47}{4}$

5. Write a fraction and a division statement for this picture.

6. These are special fractions for eighths:

   $\frac{48}{8}, \frac{56}{8}, \frac{64}{8}, \frac{72}{8}$

   a) Why do you think these fractions are special?
   Use numbers, pictures, and words to explain.
   b) Find 4 special fractions for twelfths. Show your work.

7. Write an improper fraction and a division statement for this picture.

   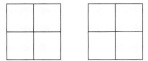

8. Draw pictures.
   Write $2\frac{2}{3}$ as an improper fraction and as a division statement.

9. Wenchun can make 4 origami swans from one sheet of paper.
   How many sheets of paper will she need to make 45 swans?

10. Jimmy has 10 rolls of string. He plans to make 7 kites.
    How much string is available for each kite?

11. Six people share a gift of $900 equally.
    How much does each person get?

12. Mario cycled 17 km from his home to visit a friend.
    He left home at 9 a.m. He arrived at his friend's home at 11 a.m.
    He cycled the same distance each hour. How far was this?

13. Janine made 5 pizzas for her party.
    She invited 7 friends.
    How much pizza did Janine think each person would eat? Explain.

Reflect

How are $\frac{13}{5}$ and $13 \div 5$ the same?
Use pictures, numbers, or words to explain.

Numbers Every Day

Number Strategies

Write each number in expanded form.
- 20.25
- 202.5
- 2.02
- 2.52

Fractions and Decimals on a Calculator

You can use a calculator to write a fraction as a decimal.

➤ Write $\frac{3}{4}$ as a decimal.

Press: 3 ÷ 4 = to display

$\frac{3}{4} = 0.75$

➤ Write $\frac{7}{5}$ as a decimal.

Press: 7 ÷ 5 = to display

$\frac{7}{5} = 1.4$

➤ $2\frac{3}{5}$ means $2 + \frac{3}{5}$.

Write $\frac{3}{5}$ as a decimal.

Press: 3 ÷ 5 = to display

Add 2.

Press: + 2 = to display

$2\frac{3}{5} = 2.6$

Practice

1. Write each fraction or mixed number as a decimal.

   a) $\frac{7}{25}$ b) $\frac{5}{4}$ c) $\frac{17}{20}$ d) $2\frac{11}{20}$ e) $3\frac{1}{4}$

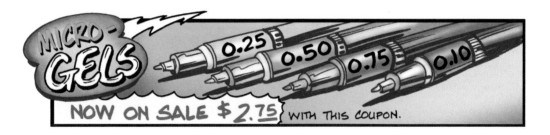

Fractions In-Between

Your teacher will give you a set of fraction and decimal cards.

- ➤ Place the decimal cards on the table in order from least to greatest.
  There should be space for 4 more cards between each pair of decimal cards.
- ➤ Shuffle the fraction cards.
  Deal 6 cards to each player.
- ➤ Take turns to place a card.
  The goal is to place the fraction cards so the fractions and decimals are in order from least to greatest.
  Place each fraction card so it touches another card.
  Fraction cards may be placed on either side of a card already on the table.
  If the fraction on the card is equivalent to one of the decimals, place the fraction directly below the decimal card.
  Cards may not be placed between two cards that are touching.
- ➤ As you place the card, name the fraction and its decimal equivalent.
- ➤ Place as many of your cards as you can.
  When neither player can place any more cards, the round is over.
  Each player receives a strike for each card left in her or his hand.
- ➤ Play 5 rounds.
  Record the strikes for each round.
  The player with the fewer strikes wins.

LESSON

7

Estimating Products and Quotients

Look at this division sentence: 63 ÷ 7 = 9
Which number is the divisor? The dividend? The quotient?

Explore

➤ A nickel has a mass of 3.95 g.
  What is the approximate mass of 7 nickels?
➤ Nine bags of dog food have
  a mass of 134.55 kg.
  What is the approximate mass of one bag?

Estimate each result.
Record your strategies and your estimates.
Show your work.

Show and Share

Share your estimates with another pair
of students.
Discuss the strategies you used to estimate.
How can you use decimal benchmarks
to estimate?

Mental Math

A veggie burger costs $3.47.
Suppose you pay with
a $5 bill.
What is the greatest
number of quarters
you could get for change?
The least number?

284 LESSON FOCUS | Estimate products and quotients with decimals.

Connect

➤ A ping-pong ball has a mass of 2.39 g.
Estimate the mass of 8 ping-pong balls.

Estimate: 2.39 × 8

2.39 is about 2.5.
Use doubles.
2 × 2.5 is 5.
Double 2 is 4: 4 × 2.5 is 10.
Double 4 is 8: 8 × 2.5 is 20.
The mass of 8 ping-pong balls is about 20 g.

> Another way to estimate is to use rounding.
>
> Round 2.39 to the nearest whole number, 2.
> 2 × 8 = 16
>
> The mass of 8 ping-pong balls is about 16 g.

➤ Four baseballs have a total mass of 575.94 g.
Estimate the mass of 1 baseball.

Estimate: 575.94 ÷ 4

Look for compatible numbers. These are pairs of numbers you can divide mentally.
575.94 is close to 600.

600 is 60 tens.
60 tens ÷ 4 = 15 tens
         = 150

The mass of 1 baseball is about 150 g.

> You can also estimate by rounding.
>
> Round 575.94 to the nearest ten, 580.
> 580 ÷ 4 = 145
>
> The mass of 1 baseball is about 145 g.

Practice

1. Estimate each product.
   For which questions did you use decimal benchmarks?
   a) 7.01 × 9
   b) 3.8 × 7
   c) 11.85 × 5
   d) 19.9 × 4
   e) 25.78 × 2
   f) 9.4 × 6
   g) 2.84 × 8
   h) 3.61 × 12

2. Estimate each quotient.
   a) 9.8 ÷ 5
   b) 12.31 ÷ 2
   c) 44.69 ÷ 9
   d) 56.09 ÷ 7
   e) 225.3 ÷ 5
   f) 44.23 ÷ 4
   g) 255.2 ÷ 5
   h) 18.29 ÷ 6

3. One football costs $14.89.
   About how much will 6 footballs cost?

4. Waldo paid $29.38 for 3 soccer balls.
   Estimate the cost of 1 soccer ball.

5. Liza drives her car to work 5 days a week.
   She drives a total of 104.6 km in one week.
   About how far does Liza drive each day?

6. Estimate the perimeter of each regular figure.
   a) 1.9 cm
   b) 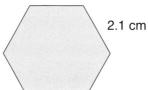 2.1 cm
   c) 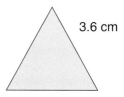 3.6 cm

7. Estimate the side length of each regular figure.
   a) Perimeter is 24.2 cm.
   b) Perimeter is 29.8 cm.
   c) Perimeter is 31.8 cm.

8. a) Is 9.47 × 5 greater than, or less than, 45?
      How can you estimate to find out?
   b) Is 23.86 ÷ 4 greater than, or less than, 6?
      How can you estimate to find out? Show your work.

9. Copy and complete. Write >, <, or =.
   How did you decide which symbol to use?
   a) 5.6 × 2 ☐ 1.4 × 4
   b) 6.8 ÷ 2 ☐ 15.5 ÷ 5

10. Write a division statement where the quotient is:
    a) between 5.5 and 10.4
    b) between 40.1 and 60.2

Reflect

Describe how to estimate the product or quotient of a decimal and a whole number. Use words and numbers to explain.

At Home

Describe a situation where you might estimate the product or quotient of a decimal and a whole number.

LESSON 8

Multiplying Decimals with Tenths

Explore

Denise drove around the track 3 times.
About how far did Denise travel?
Find the exact distance.
Use any materials that help.
Show your work.

Show and Share

Share your work with another pair of students.
How did you estimate? Calculate?
Describe the strategies you used.
How do you know your answer is correct?

Connect

Another day, Denise drove around the track 4 times.
The track is 1.6 km long.
Find how far she travelled.

Multiply: 1.6×4
Here are 3 ways to find 1.6×4.

➤ Use Base Ten Blocks
on a place-value mat.
Model 4 groups of 1.6.
Then regroup 20 tenths
as 2 ones.

LESSON FOCUS | Multiply decimals with tenths by a 1-digit whole number.   Unit 8 Lesson 8   **287**

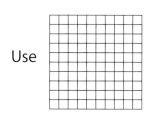

1.6 × 4 = 6.4

➤ Use grid paper.
Draw 4 groups of 1.6.

Use [ ] as 1.   Use [ ] as 0.1.

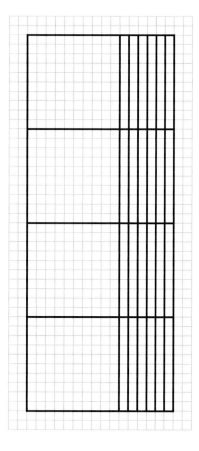

4 ones + 24 tenths
= 4 ones + 20 tenths + 4 tenths
= 4 ones + 2 ones + 4 tenths
= 6 ones + 4 tenths
= 6.4

1.6 × 4 = 6.4

➤ Use what you know about multiplying whole numbers.

```
     1.6
   ×  4
   ─────
     2.4   ←─── 4 × 6 tenths = 24 tenths or 2.4
   + 4.0   ←─── 4 × 1 = 4, or 4.0
   ─────
     6.4   ←─── 2.4 + 4.0 = 6.4
```

> 4 ones 0 tenths is the same as 4 ones. So write 4 as 4.0.

➤ Estimate to check your work.
1.6 is close to 1.5.
2 × 1.5 is 3.
4 × 1.5 is 6.
The answer will be close to 6.

Denise travelled 6.4 km.

Practice

Use grid paper or Base Ten Blocks when they help.

1. Find only the products that are greater than 40.

   a) 4.3 × 2
   b) 7.5 × 4
   c) 6.8 × 7
   d) 10.7 × 8
   e) 0.8 × 5
   f) 24.1 × 9
   g) 36.3 × 6
   h) 15.8 × 3
   i) 8.6 × 4
   j) 50.7 × 7

2. Find only the products that are less than 40.

   a) 12.2 × 8
   b) 7.9 × 5
   c) 0.6 × 9
   d) 13.1 × 2
   e) 3.6 × 7
   f) 0.9 × 9
   g) 80.4 × 6
   h) 62.3 × 7

3. Choose the number that is closest to the product. How did you choose the number?

   a) 5.9 × 8   40, 46, 47
   b) 15.7 × 3   45, 47, 50
   c) 40.2 × 5   200, 205, 210
   d) 8.1 × 4   31, 32, 33

Unit 8 Lesson 8

4. The Jonestown go-kart track is 0.9 km long.
   Suppose you went around the track 5 times.
   How far would you travel?

5. Suppose you completed 5 laps at Kendall's Kartway
   and 4 laps at the Jonestown track.
   How far would you travel in all?

6. Write a story problem you can solve by multiplying
   a decimal with tenths by a whole number.
   Solve your problem.

7. Marija multiplied 25.4 × 6.
   Her answer was 15.24.
   a) How do you know this answer is incorrect?
   b) If Marija had estimated the product,
      what would she have known?

8. Jakob has 6 gifts.
   He needs 1.4 m of ribbon to wrap each gift.
   Jakob buys 8 m of ribbon.
   Will Jakob have enough ribbon? Explain.

9. The product of two factors is 8.4.
   One factor is 1.4.
   The other factor is a 1-digit whole number.
   Find the other factor.
   Show your work.

10. Is the product of 1.23 × 8 equal to 0.984, 9.84, or 98.4?
    Without calculating, explain how you know.

Reflect

Explain why it is important to estimate
the product when you multiply a decimal
by a whole number.

Numbers Every Day

Calculator Skills

Find 3 numbers with
a product of 30 and
a sum of 9.5.

LESSON 9

Multiplying Decimals with Hundredths

Explore

The fastest growing plant in the world is a species of bamboo.
It grows at an amazing rate of 0.91 m per day.
How much does this species of bamboo grow in one week?

First estimate.
Then find ways to check your estimate.
Show your work.

Show and Share

Share your work with another pair of students.
Describe the strategies you used.
How do you know your answer is correct?

Connect

Some varieties of cactus grow only 2.54 cm per year!
How much would one of these cacti grow in 3 years?

Multiply: 2.54 × 3

Here are 2 ways to find 2.54 × 3.

➤ Use Base Ten Blocks.
   Model 3 groups of 2.54.

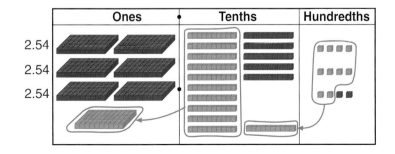

Regroup 10 hundredths as 1 tenth.
Regroup 10 tenths as 1 one.

LESSON FOCUS | Multiply decimals with hundredths by a 1-digit whole number.

Ones	Tenths	Hundredths
(7 flats)	(6 rods)	(2 small cubes)

7 ones + 6 tenths + 2 hundredths = 7.62
2.54 × 3 = 7.62

➤ Use what you know about multiplying whole numbers.

Step 1: Record the numbers without the decimal point.

Multiply as you would with whole numbers.

```
  254
×   3
-----
   12
  150
  600
-----
  762
```

Step 2: Estimate to place the decimal point in the product.

Use front-end estimation: 2.54 is about 2.
3 × 2 = 6, so place the decimal point after 7.
2.54 × 3 = 7.62

254 hundredths × 3 is 762 hundredths or 7.62

The cactus would grow 7.62 cm in 3 years.

Practice

Use Base Ten Blocks when they help.

1. Estimate. Then multiply.

   a) 2.34 × 2

   b) 6.08 × 6

   c) 3.79 × 4

   d) 5.25 × 3

   e) 7.88 × 5

   f) $4.85 × 7

2. Multiply.
   a) 9.06 × 4
   b) 7.51 × 8
   c) $3.68 × 7
   d) 4.39 × 3
   e) 0.04 × 5
   f) 8.24 × 9

3. A snail travelled 0.94 m each hour.
   How far would the snail travel in 6 h?

4. Write each amount of money using decimals.
   a) 9 quarters
   b) 7 nickels
   c) 8 fifty-cent coins

5. The Kanatas had lunch at a concession stand.
   Mrs. Kanata and her 5 children each had
   hamburgers, French fries, and water.
   How much did Mrs. Kanata pay for lunch?

6. Estimate. Is each product less than
   or greater than 15? How do you know?
   a) 2.49 × 7
   b) 3.73 × 4
   c) 5.08 × 3
   d) 8.2 × 2

7. Charlie's yard is rectangular.
   It is 20.54 m long and 9 m wide.
   What is the area of the yard?

8. Write a story problem you can solve
   by multiplying 8.75 by a whole number.
   Solve your problem.

9. Kim multiplied 8.18 × 5.
   Kim's answer was 4.09.
   What did Kim do incorrectly?

Reflect

How is multiplying a decimal by a whole number
like multiplying whole numbers?
How is it different?

Numbers Every Day

Number Strategies

Find each result.
Order the results from least
to greatest.

- 2.02 × 50
- 132.34 − 11.78
- 62.21 + 35.84
- 28.7 ÷ 2

LESSON 10

Strategies Toolkit

Explore

You will need Pattern Blocks.

Make a parallelogram that is $\frac{3}{4}$ red and $\frac{1}{4}$ blue.
Can you do this in more than one way? Explain.

Show and Share

Describe the strategy you used to solve this problem.

Connect

Use Pattern Blocks.
Make the smallest triangle you can that is $\frac{3}{16}$ green, $\frac{3}{16}$ red, $\frac{1}{4}$ blue, and $\frac{3}{8}$ yellow.

How many blocks of each colour will you need?

Understand

What do you know?
- Use Pattern Blocks to build a triangle.
- $\frac{3}{16}$ of the triangle is green.
- $\frac{3}{16}$ of the triangle is red.
- $\frac{1}{4}$ of the triangle is blue.
- $\frac{3}{8}$ of the triangle is yellow.

Plan

Think of a strategy to help you solve the problem.
- You can **use a model**.

Strategies
- Make a table.
- **Use a model.**
- Draw a diagram.
- Solve a simpler problem.
- Work backward.
- Guess and check.
- Make an organized list.
- Use a pattern.
- Draw a graph.

LESSON FOCUS | Interpret a problem and select an appropriate strategy.

Use Pattern Blocks to build the triangle.
The smallest figure is the green triangle.
$\frac{3}{16}$ of the triangle is green.
How many green triangles could you use?
How many blocks of each colour do you need to build the triangle?

Check your work.
Is $\frac{3}{16}$ of the triangle green?
Is $\frac{3}{16}$ of the triangle red?
Is $\frac{1}{4}$ of the triangle blue?
Is $\frac{3}{8}$ of the triangle yellow?

Practice

Choose one of the **Strategies**

1. Brenna can cut a log into thirds in 10 min.
   How long would it take her to cut a similar log into sixths?

2. One-fourth of a 10-m by 10-m garden is planted with corn.
   Two-tenths of the garden is planted with squash.
   Thirty-five hundredths of the garden is planted with beans.
   The rest is planted with flowers.
   What fraction of the garden is planted with flowers?
   Write your answer as a decimal.

3. A snail is trying to reach a leaf 4 m away.
   The snail crawls 2 m on the first day.
   Each day after that, it crawls one-half as far as the previous day.
   After 4 days, will the snail reach the leaf? How do you know?

Reflect

How can using a model help you to solve problems with fractions and decimals?
Use words, pictures, or numbers to explain.

Unit 8 Lesson 10

LESSON 11

Dividing Decimals with Tenths

Explore

Cecil visited his grandma in England.
While he was there, Cecil went on the *Ultimate*,
one of the world's longest roller coasters.
Cecil had 5 rides on the *Ultimate*.
He travelled a total distance of 11.5 km.
How long is the roller coaster?

Estimate first. Then solve this problem.
Use any materials you think may help.

Show and Share

Share your work with another pair of students.
Discuss the strategies you used to estimate and to check.
How do you know your answer is correct?

Connect

Mimi and 3 friends made taffy.
They stretched the taffy to a length of 13.6 cm.
When it was cool, they shared the taffy equally.
How much taffy did each person get?

Divide: 13.6 ÷ 4

Here are 2 ways to find 13.6 ÷ 4.

➤ Use Base Ten Blocks.
   Model 13.6.

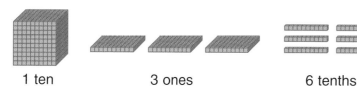

1 ten   3 ones   6 tenths

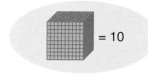 = 10

296 LESSON FOCUS | Divide decimals with tenths by a 1-digit whole number.

Trade 1 ten for 10 ones.

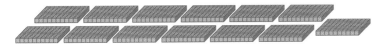

13 ones   6 tenths

Share 13 ones among 4 equal groups.

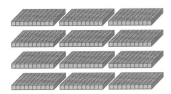 

There are 3 ones in each group,
with 1 one and 6 tenths left over.
Trade 1 one for 10 tenths. There are now 16 tenths.

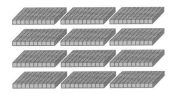 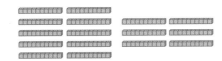

Arrange the 16 tenths among the 4 equal groups.

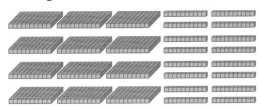

There are 3 ones and
4 tenths in each group.

$13.6 \div 4 = 3.4$

➤ Use what you know about dividing whole numbers.

Step 1: Record the numbers without the decimal point. Divide as you would with whole numbers.

$4 \overline{)1\,3\,^1 6}$
$3\;4$

Step 2: Estimate to place the decimal point.

$13.6 \div 4 = 3.4$

Check by multiplying: $3.4 \times 4 = 13.6$
Each person gets 3.4 cm of taffy.

13.6 is a little more than 12.
So, 13.6 ÷ 4 is a little more
than 12 ÷ 4 = 3.
The answer must be 3.4.

Unit 8 Lesson 11 **297**

Practice

Use Base Ten Blocks when they help.

1. Estimate each quotient.
   a) 18.7 ÷ 2   b) 28.9 ÷ 7   c) 1.9 ÷ 2   d) 150.4 ÷ 3

2. Divide.
   a) 7.4 ÷ 2   b) 9.3 ÷ 3   c) 25.6 ÷ 8   d) 85.8 ÷ 6
   e) 8.4 ÷ 6   f) 45.5 ÷ 7   g) 623.4 ÷ 3   h) 375.4 ÷ 2

3. Which number is the quotient?
   How did you choose the number?
   a) 13.5 ÷ 3   45, 4.5, 0.45
   b) 34.2 ÷ 6   5.7, 7.5, 57
   c) 78.4 ÷ 8   8.8, 98, 9.8

4. A box of 6 individual servings of oatmeal has a mass of 170.4 g.
   What is the mass of 1 serving?

5. A photo in Chelsey's book shows an insect magnified 25 times its actual size.
   In the book, the insect is 9.5 cm long.
   What is the actual length of the insect?

6. Olav walks to and from work every day from Monday to Friday.
   This is a total of 20.5 km a week.
   How far does Olav live from his workplace?
   Show your work.

7. One lap around the track is 2.7 km.
   How long is one-half of a lap around the track?
   How long is one-third of a lap?
   How do you know?

Numbers Every Day

Number Strategies

Is 31 × 42 greater than 1200?
How do you know without multiplying?

Reflect

Choose one of the questions from *Practice*.
Explain how you found the quotient.

Dividing Decimals with Hundredths

Jeremy and 4 friends earned $98.45 raking leaves last fall.
They shared their earnings equally.
How much did each person earn?

Estimate first.
Then find the amount each person earned.
Use any materials you think may help.

Show and Share

Share your solution with another pair of students.
Discuss the strategies you used to solve the problem.
How do you know your answer is correct?

Connect

Jessie rode her bicycle around the block 3 times.
She travelled a total distance of 4.92 km.
How far is it around the block?

Divide: 4.92 ÷ 3

Here are 2 ways to find 4.92 ÷ 3.

➤ Use Base Ten Blocks.
   Model 4.92.

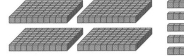 

Numbers Every Day

Number Strategies

Write 5 different number sentences with the answer 1400.

LESSON FOCUS | Divide decimals with hundredths by a 1-digit whole number.

299

Arrange the ones blocks into 3 equal rows.

 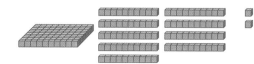

Each row has 1 one, with 1 one, 9 tenths, and 2 hundredths left over.
Trade 1 one for 10 tenths. Now there are 19 tenths.

Arrange the 19 tenths among 3 groups.
Each group has 1 one and 6 tenths, with 1 tenth and 2 hundredths left over.

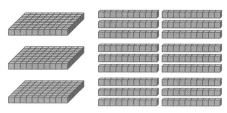 

Trade 1 tenth for 10 hundredths. Now there are 12 hundredths.
Share the hundredths blocks equally among the 3 groups.
Each group has 4 hundredths.

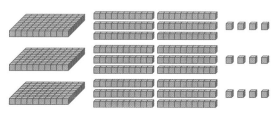

Each group has 1 one, 6 tenths, and 4 hundredths.

4.92 ÷ 3 = 1.64

▶ Use what you know about dividing whole numbers.

Step 1: Record the numbers without the decimal point. Divide as you would with whole numbers.

$$3\overline{)4^{1}9^{1}2}$$
$$\phantom{3\overline{)}}1\ 6\ 4$$

Step 2: Estimate to place the decimal point.   4.92 ÷ 3 = 1.64

4 ÷ 3 is a little more than 1. The answer must be 1.64.

Check by multiplying:
1.64 × 3 = 4.92
It is 1.64 km around the block.

Unit 8 Lesson 12

Practice

Use Base Ten Blocks when they help.

1. Divide.
   a) 4.62 ÷ 2
   b) 7.65 ÷ 9
   c) 2.04 ÷ 6
   d) 8.32 ÷ 4
   e) 26.35 ÷ 5
   f) 17.16 ÷ 6
   g) 19.14 ÷ 3
   h) 0.08 ÷ 2

2. Find only the quotients that are greater than 1.5.
   a) 7.56 ÷ 3
   b) 2.16 ÷ 9
   c) 4.38 ÷ 6
   d) 5.88 ÷ 4
   e) 8.75 ÷ 5
   f) 24.84 ÷ 2
   g) 60.32 ÷ 8
   h) 10.64 ÷ 7

3. Find the length of one side of each regular polygon.
   a) Perimeter = 11.64 m
   b) Perimeter = 18.42 m

4. A 6-pack of bottled water costs $3.54. What is the cost of one bottle?

5. a) A new tripod costs $89.46. Fourteen people will share the cost equally. How much will each person pay?
   b) The camera shop gave a $9.10 discount. How much did each person pay?
   Show your work.

6. Write a story problem that can be solved by dividing a decimal by a whole number. Solve your problem. Show your work.

7. Carl divided 4.76 by 4 and got 119. How was Carl incorrect?

Reflect

Why can you divide decimals the same way you divide whole numbers? Use words and numbers to explain.

Unit 8 Show What You Know

LESSON

1
1. How many ways can you describe the fraction of the figure that is shaded? Record each way.
   a) 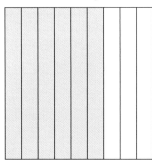   b)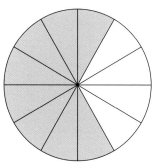

2
2. Are the fractions in each pair equivalent? How do you know?
   a) $1\frac{2}{5}$ and $\frac{7}{3}$ b) $\frac{25}{8}$ and $2\frac{5}{8}$ c) $2\frac{1}{4}$ and $\frac{10}{4}$

3
3. Order the fractions in each set from least to greatest.
   a) $\frac{12}{5}, \frac{13}{6}, \frac{14}{7}$ b) $3\frac{4}{9}, 2\frac{5}{8}, 3\frac{5}{12}$

4. A mosaic has 36 tiles.
   One-sixth of the tiles are blue,
   $\frac{5}{12}$ are green, and $\frac{2}{9}$ are tan.
   The rest are white.
   Which colour are the greatest number of tiles?

5. Which dot on the number line represents each number?
   2.3, $\frac{12}{4}$, $\frac{16}{10}$, 1.8

   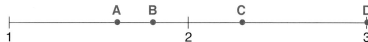

4
6. Draw a picture to represent each fraction. Then write each fraction as a decimal.
   a) $\frac{3}{5}$ b) $\frac{9}{100}$ c) $3\frac{3}{4}$ d) $1\frac{11}{25}$

7. Write 2 equivalent fractions for each decimal.
   a) 0.1   b) 0.20   c) 0.25   d) 0.9

LESSON

5   **8.** Draw a picture to represent each decimal. Then write each decimal as a fraction.
      a) 0.25   **b)** 0.6   **c)** 0.75   **d)** 0.8

6   **9.** Write each improper fraction as a mixed number.
      a) $\frac{81}{4}$ **b)** $\frac{111}{3}$ **c)** $\frac{77}{5}$ **d)** $\frac{135}{10}$

7   **10.** Estimate each product or quotient.
      a) 4.89 × 6   **b)** 19.5 ÷ 4   **c)** 2.04 × 7   **d)** 17.6 ÷ 3

11. How much would you pay for 5 snow cones?

Snow Cones $1.25 each — Buy 3, get 1 free!

8
9   **12.** Multiply.
      a) 45.6 × 2   **b)** 0.76 × 9
      c) 5.09 × 6   **d)** 8.4 × 8

11
12   **13.** Divide.
      a) 14.8 ÷ 4   **b)** 7.35 ÷ 5
      c) 12.96 ÷ 8   **d)** 5.39 ÷ 7

14. A student divided 14.63 ÷ 7 and got 20.9. Check the answer. Is it correct? How do you know?

UNIT 8 Learning Goals

- ✓ model, compare, and order fractions, improper fractions, and mixed numbers
- ✓ explore equivalent fractions and decimals
- ✓ explore patterns involving fractions
- ✓ relate fractions to division and to decimals
- ✓ estimate decimal products and quotients
- ✓ multiply decimals with tenths and with hundredths
- ✓ divide decimals with tenths
- ✓ pose and solve problems involving decimals and fractions

Unit 8

Unit Problem: In the Garden

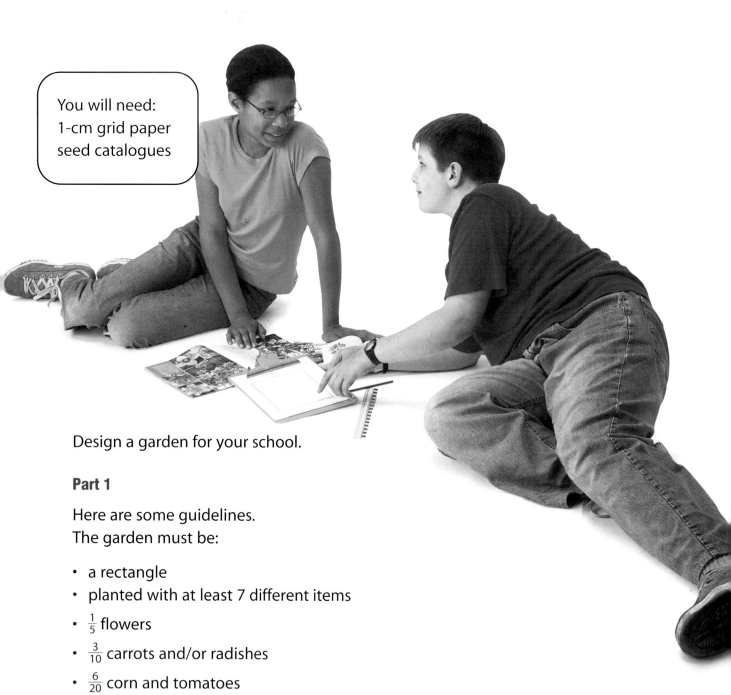

You will need:
1-cm grid paper
seed catalogues

Design a garden for your school.

Part 1

Here are some guidelines.
The garden must be:

- a rectangle
- planted with at least 7 different items
- $\frac{1}{5}$ flowers
- $\frac{3}{10}$ carrots and/or radishes
- $\frac{6}{20}$ corn and tomatoes

The tomatoes section is twice the size of the corn section.
Draw your garden on a hundredths grid.
Label each section clearly.

What fraction of the garden does each section represent?
What decimal does each section represent?

Part 2

Make up your own guidelines for designing a garden.
Exchange guidelines with another pair of classmates.
Follow the guidelines to design your classmate's garden.

Part 3

Write 2 story problems about your garden:
- One problem involves multiplication with decimals.
- The other problem involves division with decimals.

Exchange problems with another pair of classmates.
Solve your classmates' problems.
Check each other's work.

Check List

Your work should show
- ✓ a plan of the garden on grid paper, with each section clearly labelled
- ✓ the fraction or decimal each section represents
- ✓ how you calculated how to represent each section on the grid
- ✓ how you multiplied and divided decimals

Reflect on the Unit

How are fractions and decimals the same?
How are they different?

Unit 8   **305**

Units 1–8 Cumulative Review

UNIT

1

1. Write the first 6 terms of each pattern.
   a) Start at 65. Alternately subtract 8, then add 2.
   b) Start at 2. Alternately add 3, then multiply by 2.

2

2. Estimate first. Find each result greater than 2000.
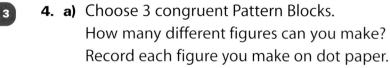
   a) 1007
       + 1088
   b) 92 × 87
   c) 7764
       − 5948
   d) 3)6496

3. Aidan plans to plant 12 rows of trees.
   There will be 11 trees in each row.
   How many trees will Aidan plant?

3

4. a) Choose 3 congruent Pattern Blocks.
      How many different figures can you make?
      Record each figure you make on dot paper.
   b) Choose 3 different Pattern Blocks.
      How many different figures can you make?
      Record each figure you make on dot paper.

4

5. Estimate first. Find each sum or difference less than 3.5.
   a) 9.4 − 3.7   b) 5.65 − 4.91
   c) 1.02 + 1.29   d) 6.3 − 5.18
   e) 4.4 + 1.72   f) 2.04 + 1.17
   g) 2.22 + 1.44   h) 8.67 − 3.54

6. One book cost $5.79. Kim had $62.48.
   She bought 10 books. How much money does Kim have left?

5

7. Here are the masses, to the nearest kilogram, of a group of dogs:
   34, 9, 27, 7, 22, 31, 31, 9, 27, 8, 14, 31, 14, 29, 24, 7, 8, 12
   a) Arrange the data into intervals.
   b) Make a frequency table.
   c) Draw a bar graph.
   d) What information can you get from the graph?

8. The table shows the value of $100 US in Canadian dollars ($ Can) on the last business day in January. Amounts are shown to the nearest dollar.

Year	Value ($ Can)
1999	151
2000	145
2001	150
2002	159
2003	152
2004	132

   a) Draw a broken-line graph to display these data.
   b) Write 2 things you know from the graph.

9. Sarah walks 10 km in 40 min and 20 km in 80 min. How long will it take Sarah to walk 50 km?

10. This object is made with centimetre cubes.
    a) Find its volume in cubic centimetres.
    b) Suppose the object is hollow. How many millilitres of water could it hold?

11. Use dot paper. Draw a figure. Label it A. Draw a second figure congruent to Figure A. Label it B. Describe the transformations that would move Figure A so it coincides with Figure B.

12. Draw a figure you think will tessellate. Trace, then cut out, 5 copies of the figure. Does the figure tessellate? If not, describe how you could change the figure so it will tessellate.

13. Order the fractions from least to greatest.
    $5\frac{3}{5}, 5\frac{3}{8}, \frac{23}{4}, 4\frac{9}{10}$

14. Find each product or quotient.
    a) 25.3×7
    b) 4.78×3
    c) $19.6 \div 8$
    d) $31.02 \div 6$

15. A student says the quotient of $54.81 \div 9$ is 60.9. Explain the student's mistake.

UNIT 9
Length, Perimeter,
At the Zoo

Learning Goals

- estimate and measure linear dimensions
- relate units of linear measure
- use decimals to report linear measures
- explore circumference
- estimate and measure perimeter and area
- relate the perimeter and area of a rectangle
- solve problems related to length, perimeter, and area

and Area

Key Words

linear dimensions

standard units

non-standard units

circumference

scale

- Which measurements can you find in this picture?
- Which measurements describe length? Height? Width? Area?
- What does "500 m by 300 m" on the property for sale sign mean?
- Do you think this property is larger or smaller than your school's property?
- How would you find the perimeter of the property for sale?
- What does 2 km² on the "Welcome" sign mean?
- Which unit would you use to measure the perimeter of the apple orchard? The perimeter of the zoo grounds? The length of the rhinoceros?

LESSON 1

Measuring Linear Dimensions

What units could you use to measure the length of this book?

Explore

You will need a ruler and a metre stick or measuring tape.

➤ Choose an object.
  Find another object that is about twice as high.
  Estimate. Then measure to the nearest unit to check.
➤ Choose a different object.
  Find another object that is about one-half as wide.
  Estimate. Then measure to check.
➤ Choose a different object.
  Find another object that is about three times as long.
  Estimate. Then measure to check.

Record your work.

Show and Share

Share your work with another pair of students.
Explain how you decided on the units to use to measure.
What strategies did you use to estimate?

Try to choose objects so you measure:
- in millimetres (mm)
- in centimetres (cm)
- in decimetres (dm)
- in metres (m)

LESSON FOCUS | Select an appropriate unit to measure linear dimensions.

Connect

When you measure the length, width, height, thickness, or depth of an object, you are measuring a **linear dimension**.

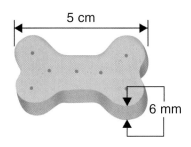

This dog biscuit is 5 cm long and 6 mm thick.

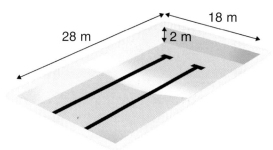

This pool is 28 m long and 18 m wide. The depth of the water in the pool is 2 m.

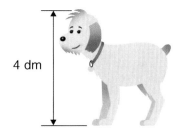

This dog is 4 dm tall.

The Hillsborough River on Prince Edward Island is 45 km long.

Practice

1. Name:
   a) an object that is about 2 mm thick
   b) an animal that is about 6 m high
   c) an object that is about 1 dm long
   d) an animal that is about 8 cm long
   e) a natural object that is about 7 m high

2. Choose the most appropriate unit for measuring each item. Explain your choice.
   a) the length of a driveway
   b) the height of a mountain
   c) the depth of a footprint in the sand
   d) the distance from Calgary to Regina
   e) the width of a baby's finger

Unit 9 Lesson 1   **311**

3. Estimate each linear dimension.
   Then measure to the nearest whole unit.
   Which tool did you use to measure? Why?
   a) the height of your desk
   b) the width of the hallway
   c) the thickness of a counter
   d) the length of a new piece of chalk

4. Draw a picture of each object. Use grid paper when it helps.
   a) a pencil 15 cm long
   b) an insect 14 mm long
   c) a book 6 cm long and 4 cm wide
   d) a flower 1 dm tall

5. Choose the better unit of length for measuring each object.
   Why is it better?
   Which tool would you use to measure each object?
   a) centimetre or metre
   b) millimetre or centimetre
   c) metre or kilometre

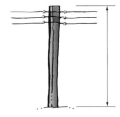  

6. Decide if each statement is reasonable.
   Explain your thinking.
   a) Ellie said that her sun parakeet has a wingspan of 15 dm.
   b) Pablo caught a salmon 25 cm long. He said the salmon was big enough to feed his family of 4.
   c) Betty says she walks 5 km to school in less than 5 min.
   d) The length of two shoelaces tied together equals the height of a small child.

Reflect

Choose an object.
Explain how you decide which unit and which tool to use for measuring its linear dimensions.

Numbers Every Day

Number Strategies

Write each number in words, and in expanded form.

21 352

210 001

LESSON 2

Relating Units of Measure

Explore

Each of you will need string, scissors, a ruler, and a metre stick or measuring tape.

➤ Cut off a piece of string you think will fit each description:
- between 1 and 2 m long
- between 50 and 100 cm long
- shorter than 1 dm

➤ Trade strings with your partner.
Measure your partner's strings to the nearest centimetre.
Then record each measurement in metres, decimetres, centimetres, and millimetres.

Show and Share

Share your measurements with your partner.
Explain how you changed centimetres to the other units of length.
How did you use decimals to record some of your measures?

Connect

There are relationships among the units you use to measure length.

➤ You can read the height of this troll in several ways.

The troll is 9 cm tall.

Since 1 cm is 0.1 dm, then 9 cm is 0.9 dm. The troll is 0.9 dm tall.

Since 1 cm is 10 mm, then 9 cm is 90 mm. The troll is 90 mm tall.

Since 1 cm is 0.01 m, then 9 cm is 0.09 m. The troll is 0.09 m tall.

LESSON FOCUS | Describe the length of an object in different units.

1 mm = 0.1 cm	1 dm = 100 mm	1 m = 1000 mm
1 mm = 0.01 dm	1 dm = 10 cm	1 m = 100 cm
	1 dm = 0.1 m	1 m = 10 dm
1 cm = 10 mm		
1 cm = 0.1 dm		1 km = 1000 m
1 cm = 0.01 m		

➤ Change 0.28 dm to millimetres.
1 dm = 100 mm
So, 0.28 dm = 0.28 × 100 mm
= 28 mm

➤ Change 3.8 dm to centimetres.
1 dm = 10 cm
So, 3.8 dm = 3.8 × 10 cm
= 38 cm

➤ Change 12 mm to centimetres.
10 mm = 1 cm
So, 12 mm = $\frac{12}{10}$ cm
= 1.2 cm

➤ Change 23 cm to metres.
100 cm = 1 m
So, 23 cm = $\frac{23}{100}$ m
= 0.23 m

Practice

Use a metre stick when it helps.

1. The Komodo dragon is the world's largest lizard.
   It can grow to a length of 3 m.
   Write this length in decimetres and in centimetres.

2. Copy and complete.
   a) 9.6 dm = ☐ cm
   b) 15 mm = ☐ cm
   c) 5.3 dm = ☐ mm
   d) 17 cm = ☐ dm
   e) 0.45 m = ☐ cm
   f) 45 cm = ☐ m

3. How many centimetre cubes do you need to make a line of each length?
   a) 50 mm
   b) 1.2 m
   c) 21.6 dm
   d) 70 mm

4. Record each measure in millimetres, decimetres, and metres.
   a) 24 cm
   b) 17 cm
   c) 80 cm
   d) 145 cm

5. Record each measure in millimetres, centimetres, and decimetres.
   a) 3 m
   b) 2.5 m
   c) 1.4 m
   d) 0.9 m

6. Draw a feather of each length.
   Then write each length in 3 different units.
   a) 50 mm b) 3 cm
   c) 1.1 dm d) 0.07 m

Number Sense

Since $0.1 = \frac{1}{10}$, we can multiply by 0.1 to change millimetres to centimetres, and decimetres to metres.

$12 \text{ mm} = 12 \times 0.1 \text{ cm} = 1.2 \text{ cm}$

7. Copy and complete. Use =, <, or >.
   Explain how you know.
   a) 5.56 m ☐ 70 dm b) 250 cm ☐ 1.46 m
   c) 16 mm ☐ 1.6 cm d) 3000 mm ☐ 2.8 m
   e) 5.3 dm ☐ 53 cm f) 2.90 m ☐ 227 cm

8. The great white shark can grow to a length of 4.9 m.
   The mako shark can grow to a length of 40 dm.
   Which shark can grow to the greater length? Explain.

9. Jackie is 123 cm tall.
   Suppose she wants to know her height in metres.
   How will the number that represents her height in metres compare to the number that represents her height in centimetres? Explain.

10. Rico is 1.21 m tall, Jeremy is 10.3 dm tall, and Sasha is 131 cm tall.
    Order the students from shortest to tallest.
    Who is tallest? By how much? Explain.

11. Hannah-Li plans to measure the width of the classroom door in millimetres and decimetres. Which will be greater: the number that represents the width in millimetres or the number that represents the width in decimetres? How do you know?

Reflect

Explain how to change a measurement from one unit to another.
Give examples to support your answer.

Numbers Every Day

Mental Math

Multiply or divide.

$27.2 \div 10$
3.08×100
$41.2 \div 10$
1.82×10
0.95×100

L E S S O N 3

Using Non-Standard Units to Estimate Lengths

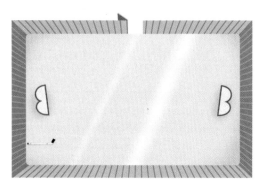

About how many hockey sticks long is the rink?

About how many car lengths are there between the blue car and the yellow car?

Explore

Suppose you do not have a ruler, a metre stick, or a measuring tape.

I estimate the hallway is 30 strides long.

➤ Choose a long length such as the length of the classroom or the width of the playground.
➤ Use a stride as a unit.
➤ Estimate the length you chose, in strides. Then measure with strides to check your estimate.
➤ Choose a new unit.
  Estimate the length you chose, in the new unit.
  Then measure to check your estimate.

Record your work.

Show and Share

Show your results to another pair of students.
Share how you chose your unit of length and how you made your estimate.
How do the measures in strides and the other unit compare?

316 LESSON FOCUS | Estimate long lengths using non-standard units.

Connect

➤ Metres and kilometres are **standard units**.
You use them to estimate and measure
long lengths and distances.

➤ Units such as floor tiles, car lengths, and strides
are **non-standard units**.
They can also be used to estimate and measure
long lengths.

➤ A car length is longer than a stride.
The measure of a distance in car lengths
will be less than the measure of the
same distance in strides.

Practice

1. Estimate each distance in strides.
   Then measure to check your estimates.
   a) the distance from the teacher's desk to the classroom door
   b) the distance from your classroom door to the principal's office

2. Compare your measurements from question 1 with those of
   two other classmates. Why are the results different?

3. Explain how you could use the height of a classmate
   to measure the length of the playground.

4. Suppose you want to find out how many tables could be
   lined up in the hallway. How could you do this
   without moving the tables?
   Are you and a classmate likely to get the
   same result? Explain.

Numbers Every Day

Calculator Skills

Add.

```
  4356
  2722
+ 5006
```

Reflect

Describe a situation when you might want to
estimate a long length using a non-standard unit.

ASSESSMENT FOCUS | Question 4

Unit 9 Lesson 3 **317**

LESSON 4

Measuring Distance Around a Circular Object

You will need a collection of circular objects, string, scissors, and a metre stick.

➤ Predict the order of the objects from the object with the greatest distance around to the object with the least distance around.
➤ Choose an object from the collection.
   Estimate the distance around the object.
   Use any materials you like to measure the distance.
➤ Record the estimate and the measurement.
➤ Repeat the activity with other objects.
➤ Order the objects from the object with the greatest distance around to the object with the least distance around.

Show *and* Share

Share your results with another pair of students.
Discuss the strategies you used to measure the distance around an object.
How did you use one measurement to help estimate the next?

Connect

The distance around a circular object or figure is its **circumference**.

Here is one way to find the circumference of a circular object. Use string and a ruler.

 Cut a length of string equal to the circumference of the object.

 Measure the string.

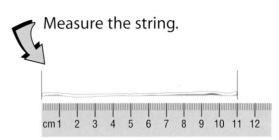

The circumference of this film canister is 11 cm.

Practice

1. Which unit and which tool would you use to measure the circumference of each item? Explain your choice.
   a) a bicycle wheel
   b) a garbage can
   c) a diamond ring
   d) a water tower
   e) a circus ring
   f) a lollipop stick

2. Estimate first. Then find the circumference of each object in the classroom.
   a) the leg of your chair
   b) a piece of chalk
   c) a globe

3. Work with a partner. Estimate, then measure, the distance around your partner's:
   a) wrist   b) baby finger   c) ankle
   Why might you want to know these measurements?

4. Find an object with each circumference. Measure to check.
   a) about 1 m
   b) about 5 cm

Numbers Every Day

Number Strategies

Find each pattern rule. Write the next 5 terms in each pattern.

• 190, 179, 168, 157, …
• 190, 192, 196, 202, …

5. You can estimate the age of a tree by measuring the circumference of its trunk.
   Each 2 cm of circumference represents about one year of growth.
   a) An oak tree has a circumference of 81 cm. About how old is the tree?
   b) Manny's elm tree is 13 years old. What is the approximate circumference of the tree?

6. Karen planted a maple key when she was 4 years old.
   Karen is now 39 years old.
   Use the data in question 5.
   What is the approximate circumference of the maple tree today?

7. A bracelet has a circumference of 15 cm.
   Could you wear the bracelet? How do you know?
   Show your work.

8. Trace a circular object.
   Look at the circle you drew. Estimate its circumference.
   How could you check your estimate?

9. Dalton plans to measure the circumference of a circular pool.
   Which will be the lesser number: the circumference in metres, or the circumference in centimetres? How do you know?

10. Use the objects from *Explore*.
    For each object, estimate about how many times as long as the width the circumference is.
    Measure the width of each object.
    Use your circumference measurements from *Explore*.
    How do you think the width and the circumference are related?

Reflect

Write a letter to a friend to explain how to find the circumference of a telephone pole.

At Home

Measure the circumference of a tree in your neighbourhood.
Use this measure to estimate the age of the tree.

LESSON 5

Using Grids to Find Perimeter and Area

Miss Dahlia likes unusual shapes. How might you find the area of her vegetable garden?

Explore

You will need an 11 by 11 geoboard, geobands, and dot paper.

➤ Make the figure to the right on the geoboard. Find its perimeter and its area.

➤ On the geoboard, make a different figure. The figure must have square corners.

You may do this: You may *not* do this:

Draw the figure on dot paper.
Find the perimeter and the area.

Show and Share

Talk with another pair of students about how you used the geoboard or dot paper to measure perimeter and area.

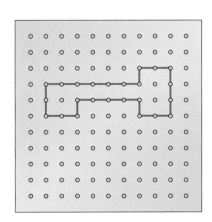

Numbers Every Day

Number Strategies

Order these numbers from greatest to least.

125 493, 215 934,
159 234, 251 439

LESSON FOCUS | Use grid paper to measure perimeter and area.

Connect

➤ One way to find the *perimeter* of this figure on 1-cm grid paper is to count the units along the outside of the figure.

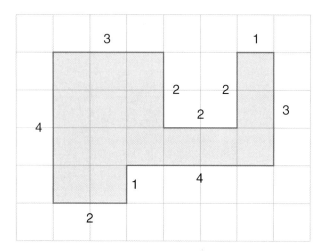

I keep track of my counting by labelling the length of each side of the figure, then adding.

Each side of every square on this grid is 1 cm long.
3 + 2 + 2 + 2 + 1 + 3 + 4 + 1 + 2 + 4 = 24
The perimeter is 24 cm.

➤ One way to find the *area* of this figure is to count the squares inside the figure.

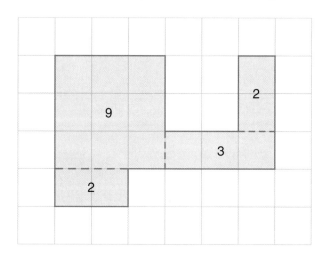

I keep track of my counting by dividing the figure into rectangles. I label each rectangle with the number of squares. Then I add.

Each square on the grid has an area of 1 cm².
9 + 2 + 3 + 2 = 16
The area of the figure is 16 cm².

Practice

1. a) Estimate first. Then find the perimeter and area of each figure on this 1-cm grid.

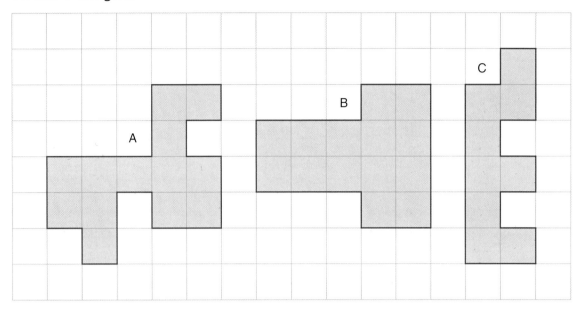

b) Order the figures from least to greatest perimeter. Then order them from least to greatest area.

2. Use 1-cm grid paper. Draw only on the lines.
Draw 3 figures.
Find and record the perimeter and area of each figure.

3. Use a geoboard. Make a figure with perimeter 16 units.
The figure must have square corners.
Draw your figure on grid paper.
  a) Explain how you know the perimeter is 16 units.
  b) What is the area of your figure?
  c) Make a different figure with perimeter 16 units.
  What is the area of this figure?

4. Use a geoboard.
Make a figure to fit each description.
Each figure should have only square corners.
  a) a perimeter of 12 units and an area of 5 square units
  b) a perimeter of 14 units and an area of 6 square units
  c) a perimeter of 10 units and an area of 5 square units
Record your work on grid paper.
Describe each figure you made.

Unit 9 Lesson 5 **323**

5. These 3 figures have different perimeters.

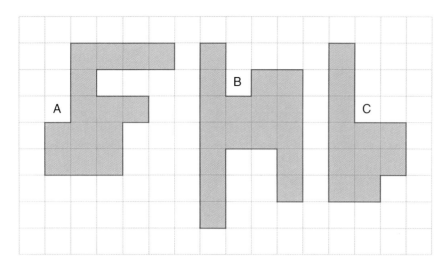

Copy the figures onto 1-cm grid paper.
Change 2 of the figures so that all 3 figures have equal perimeters.

6. Miss Dahlia wants to make a flower garden with an unusual shape. She has 28 m of fencing to enclose the garden.
   a) Draw 3 different figures on 1-cm grid paper for Miss Dahlia to choose from.
      Let the length of each square on the grid represent 1 m.
   b) Which one of your 3 figures will give Miss Dahlia the greatest amount of space for planting flowers?
   Show your work.

7. a) Use 1-cm grid paper.
      Draw a 1-cm, a 2-cm, a 3-cm, and a 4-cm square.
      Use a ruler to draw a diagonal on each square.
      Measure and record the length of each diagonal.
      Compare the side length of each square to the length of its diagonal. What do you notice?
   b) Danny said the perimeter of this figure is 10 cm. Do you agree? Explain.

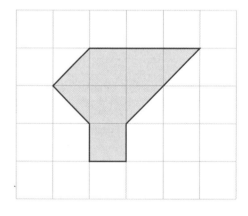

Reflect

Describe how to find the perimeter and area of a figure drawn on the lines of grid paper. Use pictures and words to explain.

LESSON 6

Measuring to Find Perimeter

 **Explore**

You will need a 15-cm by 15-cm cardboard square, a ruler, and scissors.

15 cm

➤ Draw diagonals on the square.
   Cut along the diagonals to make 4 congruent triangles.
➤ Use all 4 triangles to make a polygon.
   Sides with the same length must align.

You may do this.   You may *not* do this.

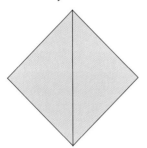 

➤ Measure to find the perimeter of the polygon.
   Record your work.
➤ Repeat with other polygons.
   Try to make one polygon with 3 sides, one with 4 sides, one with 5 sides, and one with 6 sides.

Show *and* Share

Share your work with another pair of students. Discuss the strategies you used to find the perimeters of the polygons. Did you find any shortcuts? What can you say about the area of each polygon you made?

LESSON FOCUS | Measure the perimeter of a polygon.

Connect

➤ To find the perimeter of a polygon, measure the lengths of its sides, then add.

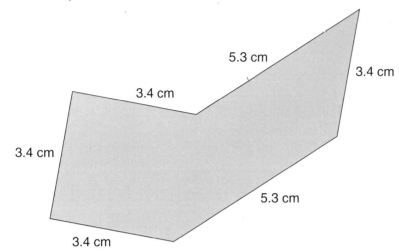

Perimeter = 5.3 cm + 3.4 cm + 5.3 cm + 3.4 cm + 3.4 cm + 3.4 cm
Perimeter = 24.2 cm

➤ Some polygons are too large to draw on a page.
A polygon like this is drawn to **scale**.
The drawing is similar to the actual polygon.
It has the same shape as the actual polygon, but it is smaller.
The length of each side is given.
To find the perimeter of the actual polygon, add the lengths of its sides.

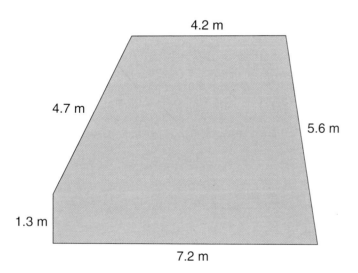

Perimeter = 4.2 m + 5.6 m + 7.2 m + 1.3 m + 4.7 m
Perimeter = 23.0 m

Numbers Every Day

Number Strategies

Estimate each sum.

1356 + 2478
5020 + 2891
4522 + 3005

Which strategies did you use?

Practice

1. Choose the most appropriate unit of length. Explain your choice.
   Measure to find the perimeter of each object.
   Which tool did you use to measure?
   a) a number key on your calculator
   b) your math book
   c) the classroom
   d) a desk or a table

2. Find the perimeter of each object.
   Write each perimeter in a different unit.

   a) 12.1 cm
   18.4 cm   18.4 cm
   12.1 cm

   b) 3.0 dm
   2.4 dm   2.4 dm
   3.0 dm

   c) 50.25 m
   30.40 m   30.40 m
   50.25 m
   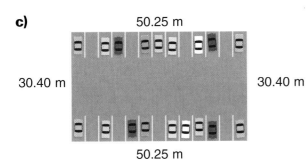

   d) 2.2 km
   0.9 km
   1.3 km
   1.8 km
   0.7 km
   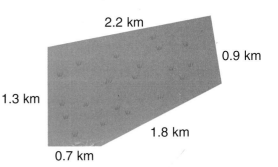

3. Estimate first. Then measure to find each perimeter.
   Write each perimeter in centimetres and in millimetres.

   a)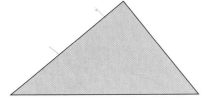

   b)

   c)

   d)

Unit 9 Lesson 6 **327**

4. Find the perimeter of each region.

a)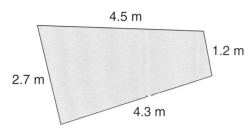

b)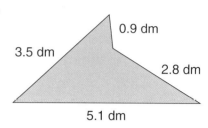

c)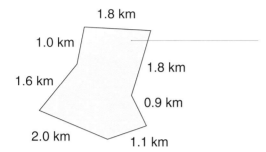

d)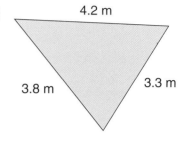

5. Find the perimeter of each figure.
   a) a rectangle with sides 1.6 dm and 2.7 dm
   b) a triangle with sides 27 mm, 16 mm, and 21 mm
   c) a pentagon with sides 1.0 m, 2.3 m, 2.2 m, 1.6 m, and 1.5 m

6. Donald's garden is rectangular. The garden is 14 m long. Its perimeter is 47 m. How wide is the garden?

7. Kim's garden is rectangular. Its perimeter is 55.0 m. Two sides of the garden have lengths between 12.0 m and 12.8 m.
   a) What might the dimensions of the garden be? How many answers can you find?
   b) How do you know the garden is not square? Show your work.

8. Which unit of length would you use to find each perimeter?
   a) a large city
   b) a swimming pool
   c) a placemat
   d) a football field
   e) a movie ticket
   f) Saskatchewan

   Explain your choice.

Reflect

Draw any polygon. Explain how to find its perimeter.

LESSON 7

Perimeter and Area

Explore

You will need 16 congruent square tiles and 1-cm grid paper.

Remember that a 2-unit by 8-unit rectangle is the same as an 8-unit by 2-unit rectangle.

➤ Use 16 tiles to make as many different rectangles as you can.
Draw each rectangle on grid paper.
Write the length, the width, the perimeter, and the area of each rectangle.

➤ Make as many different rectangles as you can that have a perimeter of 16 units.
Draw each rectangle on grid paper.
Write the length, the width, the perimeter, and the area of each rectangle.

Show and Share

Share your results with another pair of students.
What patterns do you see in your results?
How do you know you have found all the rectangles in each case?

Numbers Every Day

Number Strategies

Change the units to centimetres.

2.58 dm
258 mm
258 m

LESSON FOCUS | Relate the perimeter and the area of a rectangle.

329

Connect

➤ Rectangles with the same area can have different perimeters.
Here are 3 different rectangles with area 12 cm².

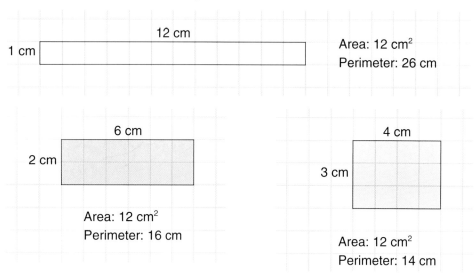

➤ Rectangles with the same perimeter can have different areas.
Here are 3 different rectangles with perimeter 12 cm.

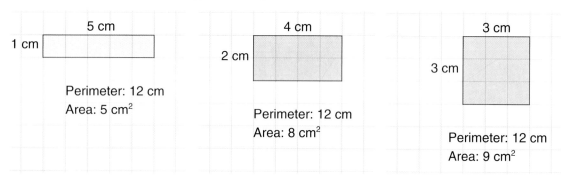

Practice

Use tiles when they help.
All the rectangles have dimensions that are whole numbers of centimetres.

1. Use 1-cm grid paper. Draw all the rectangles with each area.
   a) 8 cm²   b) 9 cm²   c) 18 cm²
   Write the length, the width, and the perimeter of each rectangle you draw.

2. Use 1-cm grid paper.
   Draw all the rectangles with each perimeter.
   a) 10 cm   b) 18 cm   c) 20 cm
   Write the length, the width, and the area of each rectangle.

3. Find the length, the width, and the perimeter of all the rectangles with each area.
   a) 5 cm²   b) 7 cm²   c) 11 cm²
   What do you notice about the results?

4. a) What is the perimeter of each rectangle?

   Rectangle A: 1 cm by 7 cm
   Rectangle B: 2 cm by 6 cm
   Rectangle C: 3 cm by 5 cm
   Rectangle D: 4 cm by 4 cm

   b) What is the area of each rectangle?
   c) What are the length and the width of the rectangle with the greatest area?

5. a) Draw 3 different rectangles, each with area 36 cm².
   b) Write the length, the width, and the perimeter of each rectangle.
   c) What are the length and the width of the rectangle with the least perimeter?

6. a) Draw a large rectangle on 1-cm grid paper.
   b) Find its area and perimeter.
   c) Draw another rectangle that has the same perimeter but is 1 cm longer. Find its length, width, and area.
   d) Draw a rectangle that has the same perimeter as the rectangle in part a, but is 1 cm shorter. Find its length, width, and area.
   e) Continue to draw rectangles with the same perimeter as the rectangle in part a. Change the length or the width by 1 cm, 2 cm, 3 cm, and so on, each time. Find the length, the width, and the area each time.
   f) What patterns do you see in the results?

ASSESSMENT FOCUS   Question 6

Math Link

Measurement and Number

When the area of a rectangle is a prime number, only 1 rectangle can be drawn with that area. The lengths of the sides are whole numbers. When the area of a rectangle is a composite number, more than 1 rectangle can be drawn with that area.

---

7. a) A rectangle has area 24 cm².
   The perimeter is to be as short as possible.
   Estimate the length and the width of the rectangle.
   Draw rectangles to check your estimate.

b) A rectangle has perimeter 24 cm.
   The area is to be as large as possible.
   Estimate the length and the width of the rectangle.
   Draw rectangles to check your estimate.

8. A rectangle has length 4 cm and width 3 cm.
   What happens to the area of the rectangle in each case?
   a) The length is doubled.
   b) The width is doubled.
   c) Both the length and the width are doubled.

9. a) Jess stated that she can draw only one rectangle if the area is a prime number.
   Is Jess correct? Explain.

b) Ronnie stated that he cannot draw a rectangle if the perimeter is an odd number.
   Is Ronnie correct? Explain.

Reflect

Can you draw more than one rectangle if the area is an odd number?
Use words and diagrams to explain.

LESSON 8

Finding the Area of an Irregular Polygon

These are regular polygons.
A regular polygon has
equal sides and equal angles.

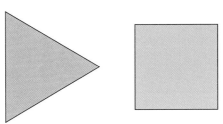

These are irregular polygons.
An irregular polygon does *not*
have all sides equal and all
angles equal.

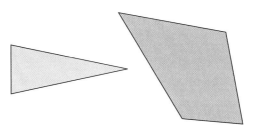

Explore

You will need a geoboard, geobands, and square dot paper.

➤ Make 4 figures:
  • an isosceles triangle
  • a trapezoid
  • an irregular pentagon
  • an irregular hexagon
➤ Draw each figure on dot paper.
  Estimate the area of each figure.
  Then find and record each area.

Show and Share

Share your figures with another pair of classmates.
Explain how you found the areas of the figures.

LESSON FOCUS | Find the area of an irregular polygon using square dot paper.

333

Connect

Here is one way to find the area of an irregular hexagon.

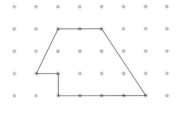

➤ Divide the hexagon into sections.

First I look for rectangles, then count squares.

➤ Find the area of the rectangle.

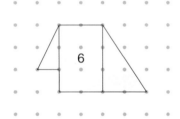

The area of the rectangle is 6 square units.

➤ Draw broken lines to make a rectangle for each triangle.
Use the area of the rectangle to figure out the area of the triangle.

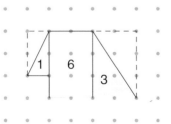

The area of one rectangle is 6 square units. The area of the triangle is $\frac{1}{2}$ the area of the rectangle. So, the area of one triangle is $\frac{1}{2}$ of 6, or 3.

➤ Add the areas of the sections to find the area of the hexagon.
1 + 6 + 3 = 10
The area of the hexagon is 10 square units.

Practice

1. Make each polygon on a geoboard. Estimate first.
   Then find the area of each figure in square units.

   a)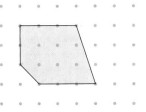

   b)

   c)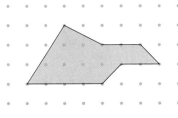

   d)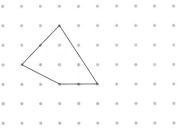

2. Order the polygons in question 1 from least to greatest area.

3. Use a geoboard.
   Make an irregular polygon with each area.
   Draw the polygon on dot paper.
   a) 5 square units
   b) 13 square units
   c) $3\frac{1}{2}$ square units
   d) 12 square units
   e) 8.5 square units
   f) 20 square units

Numbers Every Day

Number Strategies

Estimate each difference.

6048 − 3972

5856 − 4724

9147 − 6315

Which strategies did you use?

Unit 9 Lesson 8   **335**

4. Use dot paper.
   Draw 2 different polygons, each with an area of 16 square units.

5. Use a geoboard. Make each figure below.
   a) Use a geoband to divide each figure into 2 congruent parts.
      What is the area of each figure? Each congruent part?
      Record your work on dot paper.

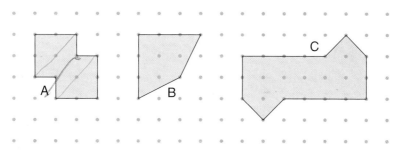

   b) Find other ways to divide each figure into
      different numbers of congruent parts.
      Find the area of each congruent part.
      Record your work on dot paper.

6. Here is a map of Gail's backyard.
   Find the area of each shaded section in square units.
   Order the sections from least to greatest area.

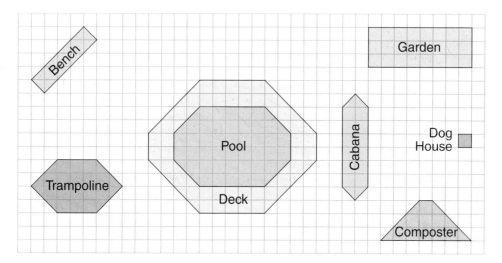

Reflect

How does using dot paper or grid paper help you to find the area of an irregular polygon? Use an example to explain.

LESSON 9

Estimating Area

Explore

You will need 1-cm grid paper.

➤ Place a shoe on the grid paper.
  Estimate how many square centimetres the shoe covers.
➤ Trace the shoe.
  Find the area of the sole of your shoe.
  Use any strategy you like.
  Record your results.

Show and Share

Show your tracing to another pair of students. Ask them to estimate the area of the tracing. Discuss the strategies you used to find the area of the sole of the shoe.
Is the area you found exact or an estimate? Explain.

Connect

Here are two ways to estimate the area of this figure:

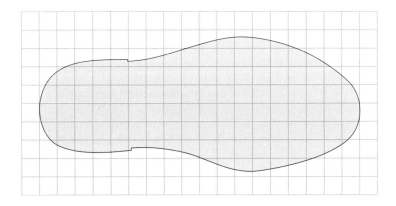

Lesson Focus | Estimate the area of a figure that is not a polygon.

➤ Draw a rectangle along grid lines inside the figure.
  Make it as big as you can. Find the area of the rectangle.
  Count the whole squares and part squares
  outside the rectangle.
  Add to find the total area.

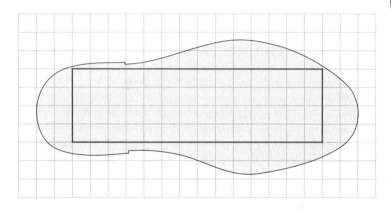

I put 2 or 3 part squares together to count as a whole square.

Area of rectangle: 14 × 4 = 56
Area outside rectangle: about 40
56 + 40 = 96
The area of the figure is about 96 square units.

I count a part larger than $\frac{1}{2}$ a square as a whole square. I ignore any part smaller than $\frac{1}{2}$ a square.

➤ Draw a rectangle along grid lines around the figure.
  Find the area of the rectangle.
  Count the whole squares and part squares
  outside the figure, and inside the rectangle.
  Subtract this number from the area of the rectangle.

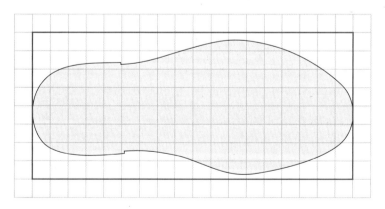

The area is approximate because it is an estimate.

Area of rectangle: 8 × 18 = 144
Area outside the figure: about 45
144 − 45 = 99
The area of the figure is about 99 square units.

Practice

1. Use 1-cm grid paper.
   a) Draw a rectangle that has about the same area as the tracing of the shoe you made in *Explore*.
   b) Draw a rectangle that has about one-half the area of the shoe tracing.

2. This is a map of several lakes.
   Each square on the map represents 1 km².

   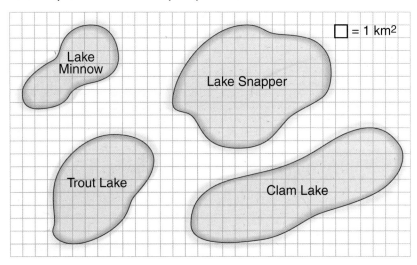

   a) Find the approximate area of each lake.
   b) Order the lakes from least to greatest area.
   c) Use grid paper. Draw a lake with an area greater than Trout Lake but less than Clam Lake.
      Find and record the area of your lake.

3. Trace a circular object on 1-cm grid paper. Find and record the circumference of the object and the area of its tracing. Explain how you found the circumference and the area. Show your work.

Reflect

Drip a few drops of water onto a paper towel. Describe how you could measure the area of the wet spot on the towel.

Numbers Every Day

Number Strategies

Order these decimals from least to greatest.

1.28, 0.37, 1.8, 0.09

LESSON 10

Strategies Toolkit

Explore

Ernesto made a 1-m square garden this year. He plans to enlarge the garden by increasing the side lengths by 2 m each year. What will the perimeter and the area of Ernesto's garden be in 6 years?

Show *and* Share

Describe the strategy you used to solve the problem.

Connect

Helen raises Angora goats.
When Helen got her first pair of goats, she built a 2-m by 1-m pen for them.
As Helen's goat population grew, she increased the size of the pen by doubling the length and the width.
What were the perimeter and area of Helen's pen after she increased its size 5 times?

Strategies
- Make a table.
- Use a model.
- Draw a diagram.
- Solve a simpler problem.
- Work backward.
- Guess and check.
- Make an organized list.
- **Use a pattern.**
- Draw a graph.

What do you know?
- Helen's first pen measured 2 m by 1 m.
- She increased the size of the pen by doubling the length and width.
- She did this 5 times.

Think of a strategy to help you solve the problem.
- You can **use a pattern**.
- Use Colour Tiles to model each pen.
- List the dimensions, the perimeter, and the area of each pen.

340 LESSON FOCUS | Interpret a problem and select an appropriate strategy.

Record your list in a table.

	Length	Width	Perimeter	Area
Original Pen	2 m	1 m	6 m	2 m²
First Increase	4 m	2 m	12 m	8 m²
Second Increase				

Look for patterns.
Continue the patterns to find
the perimeter and the area after 5 increases.

Check your work.
What pattern rules created the patterns in your table?

Practice

Choose one of the **Strategies**

1. Harold is designing a patio with congruent square concrete tiles. He has 36 tiles.
   Use grid paper to model all the possible rectangular patios Harold could build. Label the dimensions in units.
   Which patio has the greatest perimeter? The least perimeter?

2. Suppose you have a 7-cm by 5-cm rectangle.
   You increase the length by 1 cm and decrease the width by 1 cm.
   You continue to do this.
   What happens to the perimeter of the rectangle? The area?
   Explain why this happens.

Reflect

How does using a pattern help you solve a problem?
Use pictures, words, or numbers to explain.

Unit 9 Show What You Know

LESSON

1, 2

1. Measure one dimension of each object below to the nearest unit.
   Which tool did you use?
   Record each measurement using as many units as you can.
   a) a pencil case b) a stapler c) a computer screen d) a table

2. Draw a line 1.6 dm long.
   Write the measurement using as many different units as you can.

2

3. Copy each statement. Use =, >, or < to make the statement true.
   a) 1.35 m ☐ 14.3 dm
   b) 48 mm ☐ 3.7 cm
   c) 75 cm ☐ 7.5 dm
   d) 2 km ☐ 1367 m
   e) 267 cm ☐ 2.67 m
   f) 895 mm ☐ 8.98 m

4. Can you walk 100 000 mm in 2 min? Explain.

3

5. Estimate first. Then measure the length of the blackboard using a non-standard unit. Record your estimate and your measurement.

4

6. Find a cylindrical object such as a soup can.
   Which do you think is longer—the height of the can or its circumference?
   Measure to check your prediction. What did you find?

5, 6

7. Find the perimeter of each figure.
   Explain how you found each perimeter.
   Write each perimeter in a different unit.

   a)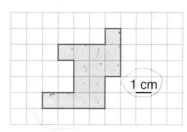

   b)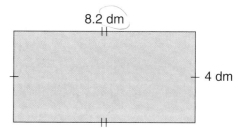

   c)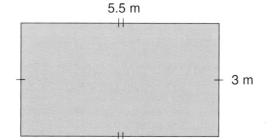

   d)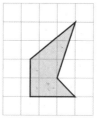

LESSON

8. Find the area of each figure in question 7.
   Explain how you did this.
   For which figures is your measure of area an estimate? Explain.

9. Use Colour Tiles. Make this pentomino:
   a) Add tiles to the pentomino to make a figure with perimeter 18 units. Draw your figure on grid paper.
   b) What are the fewest tiles you can add to the pentomino to make a figure with perimeter 18 units? Draw your figure.
   c) What are the most tiles you can add to the pentomino to make a figure with perimeter 18 units? Draw your figure.
   d) Find the area of each figure you drew in parts a, b, and c.

10. a) A rectangle has perimeter 36 cm.
    The length and the width are whole numbers of centimetres.
    The area is to be as large as possible.
    What are the length and the width of this rectangle?
    b) A rectangle has area 36 cm².
    The length and the width are whole numbers of centimetres.
    The perimeter is to be as large as possible.
    What are the length and the width of this rectangle?

11. Darlene built a rectangular dog pen with area 24.5 m². The length of the pen is 5 m. What is the width of the pen?

12. The building is one storey and rectangular.

    What might the dimensions of the floor be?
    Give 3 different answers.

UNIT 9 Learning Goals

- ✓ estimate and measure linear dimensions
- ✓ relate units of linear measure
- ✓ use decimals to report linear measures
- ✓ explore circumference
- ✓ estimate and measure perimeter and area
- ✓ relate the perimeter and area of a rectangle
- ✓ solve problems related to length, perimeter, and area

Unit 9 **343**

Unit Problem: At the Zoo

Check List

Your work should show
- ☑ a map of the petting zoo on grid paper, with each section outlined and labelled
- ☑ the dimensions, perimeter, and area of each section and how you found them
- ☑ a different shape for each region
- ☑ that the size of a region reflects the size of the animal

Reflect on the Unit

How are linear dimensions, perimeter, and area related? Write what you know about them.

UNIT 10
Patterns in Number

Squares Everywhere!

Learning Goals

- use patterns to solve multiplication problems
- model patterns in tables and on graphs
- explore tiling patterns
- use a computer to create a tiling pattern

and Geometry

Key Words

Fibonacci sequence

Fibonacci number

What patterns do you see?

How are the patterns the same?

How are they different?

How could you extend any of the patterns?

LESSON 1

Patterns in Multiplication

Explore

You will need a calculator.

➤ Predict each product.
- 4 × 5 × 18
- 4 × 18 × 5
- 2 × 36 × 5
- 18 × 5 × 4
- 18 × 10 × 2
- 9 × 10 × 4
- 9 × 10 × 2 × 2

Use a calculator. Find each product.
What do you notice about the products?
Compare the factors in the questions.
How are they related?

Before I do the next calculation, I press ON/C to clear the memory.

➤ Choose 3 different 1- or 2-digit numbers.
Multiply the 3 numbers in as many orders as you can.
How many arrangements can you make?
Which order makes multiplication easiest?

➤ Predict each product.
- 5 × 21 × 10
- 2 × 12 × 30
- 4 × 5 × 20

Use a calculator. Find each product.
How do the products compare to your predictions?

➤ Use a calculator.
Find each missing factor.
- □ × 15 = 240
- 12 × □ = 240

What strategies did you use?

348 LESSON FOCUS | Use patterns to solve multiplication problems.

Show and Share

Share your answers with another pair of students.
How did you predict each product?
Describe any patterns.
Share your strategies for finding the missing factor.

The order in which you multiply factors does not matter.
You can rearrange factors to make multiplication easier.

 Multiply: 3 × 200 × 14

The product 3 × 200 × 14 is the same as
the product 3 × 14 × 200.

Multiply: 3 × 14 = 42
Use place value to multiply 42 × 200.
42 × 2 hundreds = 84 hundreds
42 × 200 = 8400

3 × 200 × 14 = 8400

 Multiply: 5 × 12 × 3

Rewrite 12 as 2 × 6.
5 × 2 × 6 × 3
Multiply: 5 × 2 = 10
Multiply: 6 × 3 = 18

Then multiply: 10 × 18 = 180

5 × 12 × 3 = 180

12 is the product of 2 and 6.

 Multiply 12 × 29.

The product 12 × 29 is 12 less than the product 12 × 30.
Use mental math to find the product 12 × 30.
12 × 3 tens = 36 tens
12 × 30 = 360

360 − 12 = 348

12 × 29 = 348

Unit 10 Lesson 1

➤ Here are 2 ways to find a missing factor.

- ☐ × 5 = 45

  If the numbers are small, think about the facts you know.
  9 × 5 = 45
  So, the missing factor is 9.

- 285 = 15 × ☐

  When the numbers are large, use a calculator.

  Think: Which number do we multiply 15 by to get 285?

  This is the same number we get if we divide:
  285 ÷ 15
  Use a calculator.
  285 ÷ 15 = 19

  So, 285 = 15 × 19
  The missing factor is 19.

Practice

1. Use mental math to multiply. What strategies did you use?
   - a) 2 × 16 × 5
   - b) 25 × 5 × 2
   - c) 2 × 125 × 5
   - d) 2 × 34 × 50
   - e) 3 × 15 × 10
   - f) 4 × 40 × 6

2. a) Use mental math to multiply 15 × 20.
      Use this product to find:
      15 × 19   15 × 21
   b) Draw diagrams to show why the results in part a are correct.
   c) What other products could you find when you know 15 × 20? Explain.

3. Multiply 10 × 25.
   Use the product 10 × 25 to find the missing factors.
   - a) ☐ × 25 = 275
   - b) 25 × ☐ = 225
   - c) 200 = ☐ × 25
   - d) 300 = 25 × ☐

4. Find each missing factor.
   - a) ☐ × 9 = 72
   - b) 56 = 8 × ☐
   - c) 11 × ☐ = 121
   - d) 108 = ☐ × 12

5. Find each missing factor.
   a) ☐ × 50 = 1250
   b) 40 × ☐ = 600
   c) 672 = 28 × ☐
   d) 882 = ☐ × 21

6. Make up your own missing factor problem.
   Trade problems with a classmate.
   Solve your classmate's problem.

7. Use a calculator to explore the pattern.
   Copy and complete the pattern to find the missing factors.
   99 × 9 = 891
   99 × 99 = 9801
   99 × 999 = 98 901
   99 × ☐ = 989 901
   99 × ☐ = 9 899 901

8. Abdi wants to find the area of the gym floor.
   The dimensions are 33 m by 30 m.
   How can Abdi use patterns and mental math to find the area?
   Use words and numbers to explain.

9. Emma raised $4200 for charity.
   She cycled in a bike-a-thon.
   Emma's sponsors paid her $56 per kilometre.
   a) Find how far Emma cycled.
   b) How can you write this problem as a missing factor problem?

10. How could you use mental math to multiply 2 × 7 × 5 × 8?

Reflect

How can using patterns help you to multiply?
Use words and numbers to explain.

Numbers Every Day

Calculator Skills

Copy and complete each number pattern. Write each pattern rule.

3.75, 7.50, 11.25, ☐, ☐, ☐

2.4, 7.2, 21.6, 64.8, ☐, ☐, ☐

2.25, ☐, 9.25, 12.75, ☐, ☐

ASSESSMENT FOCUS | Question 8

LESSON 2

Graphing Patterns

Explore

You will need congruent squares and grid paper.
Each congruent square has sides 1 unit long.

➤ Use congruent squares to build a larger square.
  Find the perimeter of the large square.
  Record the side length and the perimeter in a table.

➤ Build 5 different squares.
  Record the side length and
  the perimeter of each square.

➤ Look at the table.
  What happens to the perimeter
  of a square as the side length
  increases? How do you know?

➤ Your teacher will give you a copy
  of this grid.
  Draw a broken-line graph to display
  the data in your table.

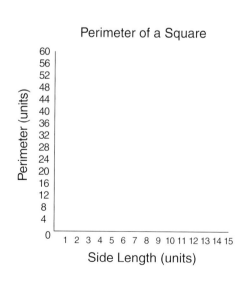

352 LESSON FOCUS | Use models, tables, and graphs to represent patterns.

Show and Share

Share your work with another pair of students.
Describe any patterns in your table.

Connect

The table shows the dimensions
of some rectangles with perimeter 52 cm.

Length (cm)	24	20	16	12
Width (cm)	2	6	10	14

Draw a broken-line graph to display
these data.

➤ Draw 2 axes.
   Label the horizontal axis "Length" and the vertical axis "Width."
   Give the graph a title.
➤ Use the same scale on both axes.
   The scale is 1 square represents 2 cm.
➤ Mark a point at *Length* 24 cm and *Width* 2 cm.
➤ Mark points for the rest of the data in the same way.
   Use a ruler to connect the points.

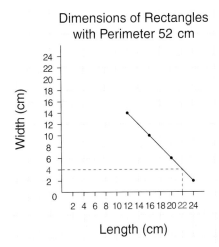

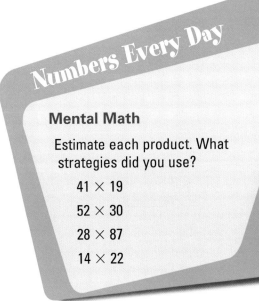

Mental Math

Estimate each product. What strategies did you use?

41 × 19
52 × 30
28 × 87
14 × 22

From the graph, the length of a rectangle
with perimeter 52 cm and width 4 cm is 22 cm.

Unit 10 Lesson 2

1. Juanita is paid $6 an hour to mow lawns.
   The table shows her earnings.
   a) Write the pattern rule for the amount earned.
   b) Draw a broken-line graph to display these data.
      How did you decide on the scale?
   c) Suppose Juanita works 3 h.
      Use the graph to find how much she will earn.
      How could you use the table to find
      how much she will earn?

Hours	Amount Earned ($)
2	12
4	24
6	36
8	48

2. Here is a growing pattern in the perimeters
   of figures made with regular hexagons.

     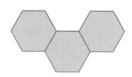

   Frame 1   Frame 2   Frame 3

Frame	Perimeter (units)
1	6

   The side length of each hexagon is 1 unit.
   a) Copy and complete the table for the first 5 frames.
   b) Predict the perimeter of the figure in Frame 6.
      How did you make your prediction?
   c) Draw a broken-line graph to display these data.

3. Use congruent squares.
   Build this growing pattern.

   Frame 1   Frame 2   Frame 3

   a) Build Frame 4 and Frame 5.
   b) Find the area of the figure in each frame.
      Record the frame numbers and the areas in a table.
   c) Predict the area of the figure in Frame 8.
   d) Use the data in the table.
      Draw a broken-line graph.

4. Copy and complete the table for this Input/Output machine.

Input	Output
2	
4	
6	
8	
10	

   a) Write the pattern rules for the input numbers and the output numbers.
   b) Use the data in the table. Draw a broken-line graph.
   c) How could you use the graph to find the output, when the input is 7?
   d) How could you use the graph to find the input, when the output is 14?

5. Use the Input/Output machine and your table from question 4.
   a) Suppose the input numbers doubled. Predict the new output numbers. How do you think the graph will change?
   b) Double the input numbers. Find the new output numbers.
   c) Graph the new input/output numbers. How does this graph compare to your graph from question 4?

6. Draw an Input/Output machine. Choose a number and an operation.
   a) Choose 5 input numbers that are multiples of any number. Find the output numbers. Make an Input/Output table for the results.
   b) Draw a broken-line graph of the data in the table. What can you find out from your graph?

Reflect

Do you prefer to display a pattern in a table or a graph? Explain your choice.

ASSESSMENT FOCUS | Question 6

Unit 10 Lesson 2 **355**

LESSON 3

Another Number Pattern

Explore

➤ Look at these number patterns:
- 2, 2, 4, 6, 10, 16, 26, …
- 5, 5, 10, 15, 25, 40, …
- 3, 3, 6, 9, 15, 24, 39, …

Identify each pattern rule.
Write the next 2 terms in each pattern.

➤ Make up a similar pattern.
Trade patterns with your partner.
Write the rule for your partner's pattern.

Show and Share

Share your work with another pair of students.
How are these number patterns different from other number patterns you have seen?

Connect

Here is a similar number pattern.
1, 1, 2, 3, 5, 8, 13, 21, …

This number pattern is called the **Fibonacci sequence**.
It is named for the Italian mathematician, Leonardo of Pisa, who was known as *Fibonacci*.

The first 2 terms of the Fibonacci sequence are 1 and 1.
After the first 2 terms, each term is the sum of the previous 2 terms.

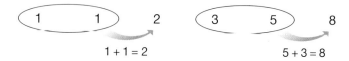

356 LESSON FOCUS | Explore number patterns in context.

Any term in the Fibonacci sequence is called a **Fibonacci number**.
Fibonacci numbers are frequently found in nature.

Practice

1. **a)** Find the 13th Fibonacci number.
   b) What is the sum of the first 13 Fibonacci numbers?

2. **a)** Write the first 16 Fibonacci numbers.
   b) Choose 3 consecutive Fibonacci numbers.
   Find the product of the least and greatest numbers.
   Multiply the middle number by itself.
   What do you notice about the products?
   c) Repeat part b for 3 different consecutive
   Fibonacci numbers. What do you notice?

3. **a)** Find each product:
   2×2 5×5
   b) Find the difference of the products in part a.
   What is special about 2, 5, and your answer to part a?

4. Repeat question 3 using the numbers 3 and 8.
   What do you notice?
   Do you think this is always true? Explain.

5. There are 2 spiral patterns on pine cones.
   The number of spirals to the left and the number of
   spirals to the right are consecutive Fibonacci numbers.
   Suppose 1 spiral pattern has 21 spirals.
   How many spirals could be in the other spiral pattern?
   How do you know?

Math Link

Nature

The number of petals on a flower is often a Fibonacci number.

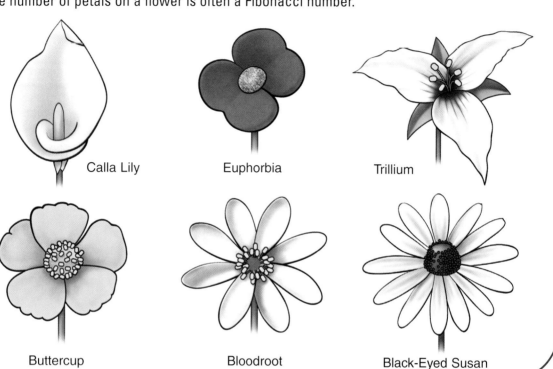

Calla Lily   Euphorbia   Trillium

Buttercup   Bloodroot   Black-Eyed Susan

6. You will need dominoes.
   The length of a domino is 2 units. Its width is 1 unit.
   a) Suppose you want to make rectangles with one side 2 units long.
      How many different rectangles can you build using 1 domino? 2 dominoes? 3 dominoes? 4 dominoes?
   b) Predict the number of different rectangles you could build using 6 dominoes.
      How could you check your prediction?
   c) How do your results relate to the Fibonacci numbers?

Reflect

Describe how to create the Fibonacci sequence. Use numbers and words to explain.

Numbers Every Day

Mental Math

Find each product.
 $4 \times 3 \times 5$
 $5 \times 7 \times 2$
 $2 \times 2 \times 2 \times 2$
 $6 \times 3 \times 5$

Which strategies did you use?

Choreographer

When you watch a music video or theatre production that has dancing, you are watching the art of a choreographer. A choreographer arranges patterns of dance steps in a performance. The dance makes the music come alive.

The choreographer must consider many things:
- the music
- the abilities of the performers
- the space available for the performance
- both group patterns and solo patterns
- the lighting
- the camera angle

Choreography is not limited to dancing, music videos, or movies. *Step and Stomp*, *Feet on Fire*, and *Funky Knees* are not new dances. They are patterns of movement designed by aerobic instructors to exercise the body. Fitness activities can be more effective and more fun when the pattern is right!

Over the years, choreographers have developed different systems for recording dance steps and other movement patterns. How do you think these dance steps could be recorded?

LESSON 4

Strategies Toolkit

Explore

A single human cell divides to form 2 new cells.
Each new cell divides in 2.
This process continues.
Suppose you start with a single human cell.
How many cells will there be after 8 rounds of division?

Show and Share

Describe the strategy you used to solve the problem.

Connect

Suppose a cow produces her first female calf when she is 2 years of age.
After that, she produces a female calf each year.
Suppose each calf produces her first female calf when she is 2 years of age and no cows die.
How many cows are there after 5 years?

Strategies
- Make a table.
- Use a model.
- **Draw a diagram.**
- Solve a simpler problem.
- Work backward.
- Guess and check.
- Make an organized list.
- Use a pattern.
- Draw a graph.

What do you know?
- Each cow produces a female calf at age 2.
- Every year after that, she produces 1 female calf.
- No cows die.

Think of a strategy to help you solve the problem.
- You can **draw a diagram**.
- Find out how many cows there are after 1 year, then after 2 years, and so on.

360 LESSON FOCUS | Interpret a problem and select an appropriate strategy.

Copy and continue the diagram.

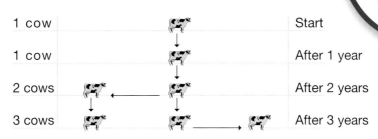

After 1 year, there is 1 cow.
After 2 years, there are 2 cows.
After 3 years, there are 3 cows.
How many cows are there after 5 years?

Check your work.
What pattern do you see in the numbers of cows?

Practice

Choose one of the **Strategies**

1. A mouse crawls through this maze. The mouse always moves forward.

   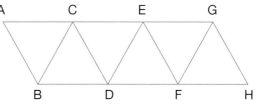

   a) How many different paths could the mouse take from A to B? From A to C? From A to D? What pattern do you see?
   b) Predict the number of different paths the mouse could take from A to H.

2. Here is a regular pentagon. Copy the pentagon. Draw all its diagonals.
   How many different triangles are there?
   How many of each type are there?

Reflect

How does drawing a diagram help to solve a problem? Use words, pictures, and numbers to explain.

Unit 10 Lesson 4 **361**

LESSON 5

Tiling Patterns

Here is a tiling pattern. It has no gaps or overlaps.

Describe some tiling patterns you see in your classroom.

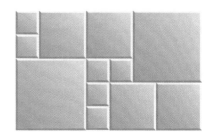

Explore

You will need Pattern Blocks and triangular dot paper.

➤ Choose 2 colours of Pattern Blocks.
  Try to arrange the blocks in a tiling pattern.
  Record the pattern.
➤ Choose 3 colours of Pattern Blocks.
  Try to arrange the blocks in a tiling pattern.
  Record the pattern.
➤ Look at each pattern you recorded.
  Describe the smallest part of the pattern that repeats.

Show and Share

Share your arrangements with another pair of students.
Which combinations of blocks covered a surface with no gaps or overlaps?
Which did not?
Describe your tiling patterns.

Numbers Every Day

Number Strategies

Order the numbers in each set from least to greatest.

- 8.5, $7\frac{1}{4}$, 7.5, $8\frac{3}{4}$
- 6.2, 7.5, $7\frac{2}{10}$, $6\frac{7}{10}$
- 3.25, $2\frac{3}{4}$, 2.25, $3\frac{1}{2}$
- 2.5, 1.5, $\frac{8}{10}$, $\frac{18}{10}$

LESSON FOCUS | Model and extend tiling patterns.

Connect

This tiling pattern is made with irregular pentagons.

The pattern uses these pentagons:

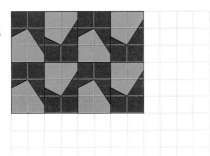

➤ To extend the pattern, continue to add pentagons. Arrange a pair of pentagons to form a rectangle.

Add the rectangle to the pattern.
Continue until the pattern is the size you want.

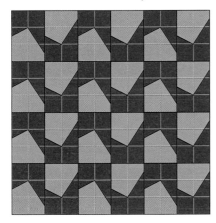

➤ The rectangle formed by a pair of pentagons is 2 units wide and 3 units long.
Suppose there are 25 pentagons of each colour. Are there enough pentagons to cover a surface that is 12 units by 12 units? How do you know? Look at the picture above.

6 rectangles fit across. 4 rectangles fit down.
Each rectangle has 1 pentagon of each colour.
6 × 4 = 24
24 pentagons of each colour are needed to cover a surface that is 12 units by 12 units.
25 is greater than 24.
There are enough pentagons to cover the surface.

Practice

1. Here is the beginning of a tiling pattern.
   a) Identify the figures in the pattern.
   b) Copy the pattern on grid paper. Extend the pattern 1 more row.
   c) The distance between grid lines is 1 unit. How many of each figure do you need to tile a surface that is 18 units by 24 units? How do you know?

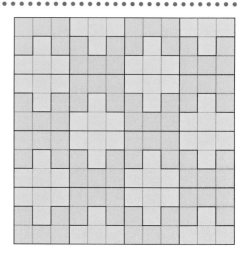

2. For each pattern, describe:
   • the figures in the pattern
   • the part of the pattern that repeats
   • how you could extend the pattern

   a) 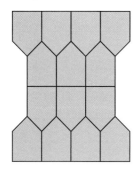   b)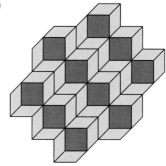

3. Use square dot paper. Create a tiling pattern with more than one figure. Describe your pattern.

Reflect

Describe how to use 2 figures to create a tiling pattern. Use pictures and words to explain.

Look for tiling patterns at home. Describe the patterns you see.

Using a Computer to Explore Tiling Patterns

Work with a partner.

Use *AppleWorks*.
Follow these steps to explore tiling patterns on a computer.

1. Open a new drawing document. Click:

2. If a grid appears on the screen, go to Step 3.

   If not, click: Options , then click: Show Graphics Grid

3. Check the ruler units are centimetres.

   Click: Format

   Click: Rulers ▶

   Then click: Ruler Settings...

   Choose these settings:

   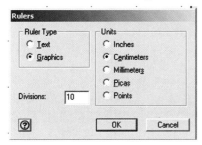

   Click: OK

4. Use these tools to draw:

   To make a **rectangle**, click the Rectangle Tool.
   Click and hold down the mouse button.
   Drag the cursor until the rectangle is the size and shape you want.
   Release the mouse button.

   To make a **square**, hold down the Shift key while you click and drag.

   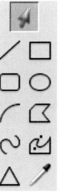

To make an **irregular polygon**, select
the Polygon Tool.
Click and drag to make each side of the polygon.
Double-click when you have finished.

To make a **regular polygon**, select the Regular Polygon Tool.

Click: Edit , then click: Polygon Sides...

Type in the number of sides you want.

Click: OK

5. To **colour a figure**, click the figure to select it.
   Click the Fill formatting button:
   Click the Color palette button:
   Then select a colour.

6. To **translate a figure**, put the cursor inside the figure.
   Click and hold down the mouse button.
   Drag the figure to where you want it.
   Release the mouse button.

7. To **copy a figure**, select the figure.

   Click: Edit , then click: Copy   Ctrl+C

   Click: Edit , then click: Paste Ctrl+V

   The copy shows on top of the figure.
   Click and drag the copy to where you want it.

8. To **reflect a figure**, select the figure.

   Click: Arrange

   Click:  or

   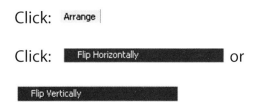

9. To **rotate a figure**, select the figure.

   Click: Arrange

   Click: Free Rotate   Shft+Ctrl+R

   Put the cursor on one of the black squares on the edge of the figure.
   Click, hold down the mouse button, and drag the figure until it is in the position you want.

   To **rotate a figure through a quarter turn**, select the figure.

   Click: Arrange

   Click: Rotate 90 Degrees

10. Use Steps 4 to 9 to create a tiling pattern.

11. Save your pattern.

    Click: File, then click: Save As...   Shft+Ctrl+S

    Name your file. Click: Save

12. Print your pattern.

    Click: File, then click: Print...   Ctrl+P

    Click: OK

Reflect

Why would a designer use a computer to create tiling patterns? Explain.

Unit 10 Show What You Know

LESSON

1

1. Use mental math to multiply. What strategies did you use?
   a) $2 \times 14 \times 5$
   b) $23 \times 5 \times 2$
   c) $2 \times 150 \times 5$
   d) $2 \times 44 \times 10$
   e) $3 \times 25 \times 10$
   f) $5 \times 17 \times 20$

2. Multiply 10×15.
   Use the product 10×15 to find the missing factor.
   a) $15 \times \square = 165$
   b) $\square \times 15 = 135$
   c) $180 = \square \times 15$
   d) $15 \times \square = 120$

2

3. You will need Colour Tiles.
   Here is a pattern with Colour Tiles.
   Each frame has one more layer than the frame before.
   Each frame has one more tile added to each "arm" than the frame before.

    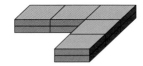 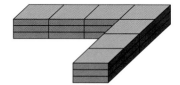

   Frame 1   Frame 2   Frame 3

   a) Build Frame 4 and Frame 5.
   b) Find the number of tiles in each frame.
      Copy and complete this table.

Frame Number	Number of Tiles
1	3

   c) Predict the number of tiles in Frame 6.
      How did you make your prediction?
      Build Frame 6 to check.
   d) Use the data in your table.
      Draw a broken-line graph.

4. Use question 3 as an example.
   Use Colour Tiles.
   Build your own 3-dimensional pattern.
   Describe your pattern.

LESSON 2

5. The pendulum on a grandfather clock swings back and forth 60 times each minute.
The table shows the number of swings.

Time (min)	Number of Swings
1	60
2	120

a) Copy and complete the table for the first 5 min.
b) Write the pattern rule for the number of swings.
c) Use the data in the table. Draw a broken-line graph.

LESSON 3

6. Pineapple scales show 3 sets of spirals.

One set goes up to the left.

A second set goes up to the right.

A third set goes up steeply to the left.

The numbers of spirals are consecutive Fibonacci numbers.
The spiral pattern that goes up steeply to the left has the greatest number of spirals, 13.
How many spirals are in the other 2 spiral patterns? How do you know?

LESSON 5

7. Devon has these Pattern Blocks:
- orange squares
- yellow hexagons
- blue rhombuses

Can he use all 3 types of blocks to create a tiling pattern?
Use words and pictures to explain.

UNIT 10 Learning Goals
- ✓ use patterns to solve multiplication problems
- ✓ model patterns in tables and on graphs
- ✓ explore tiling patterns
- ✓ use a computer to create a tiling pattern

Unit 10 **369**

Unit Problem

Squares Everywhere!

You will need 1-cm grid paper.

Part 1

On grid paper, draw a square with each side length:
1 cm, 2 cm, 3 cm, up to 8 cm

Find the area of each square.
Record the side lengths and the areas in a table.

How is the area of a square related to
its side length?
Predict the area of a square with
side length 12 cm.

Graph the data in your table.
Do not connect the points.

Part 2

On grid paper, draw a square with side length 16 cm.
Mark a point at the middle of each side.
Connect the points to make a smaller square.
Continue until you have a square with side length 2 cm.

How many squares did you draw?

Look at the squares with sides on grid lines.
Label each square with a different letter.

Record the side lengths and the areas of the squares in a table.
What do you notice? Explain.

Part 3

Create your own pattern with squares.
Colour your pattern to show a design.

Check List

Your work should show:
- ✓ the squares you are asked for clearly drawn and labelled on grid paper
- ✓ tables with the lengths of the sides and the area of each square
- ✓ clear explanations of your findings and predictions
- ✓ different ways of representing a pattern you create with squares

Reflect on the Unit

Write about the different ways you can represent patterns.
Use words, pictures, tables, and numbers to explain.

Key Words

certain

likely

outcome

likelihood

unlikely

probable

improbable

impossible

equally likely

experiment

fair game

- Is it more likely that the secret fish is silver or pink? Why?
- Tyler picks a small fish without looking. How likely is it that the fish is blue? Red? Yellow? Green?
- How likely is it that the next customer will win the free gift?

LESSON 1

The Likelihood of Events

How will you know what to wear when you leave the house tomorrow?

You cannot be **certain** of the weather. In each season, some weather conditions are more **likely** than others.

Explore

You will need a class list.

➤ Cut out each name.
  Place the names in a paper bag.
  Take turns drawing a name from the bag without looking.
  Replace the name before the next student draws.

➤ How likely is each **outcome**?
  Use these words to describe each outcome: impossible, unlikely, likely, certain
  • Someone in your class is chosen.
  • The student chosen is in grade 5.
  • Your name is chosen.
  • The student chosen is not in your class.
  • The name of the student chosen begins with a consonant.

➤ Conduct the experiment 20 times.

➤ Record your results.
  Compare the results with your predictions.

Numbers Every Day

Number Strategies

Estimate each sum.
Which strategies did you use?

1254 + 3861

3868 + 4444

6514 + 2022

2901 + 3101

374 LESSON FOCUS | Use the language of probability to describe events.

Show and Share

Compare your results with those of another group of students. Are they the same? How did you decide how likely each outcome was? Can you use a different word to describe the likelihood of any of the outcomes? Explain.

> **Likelihood** means how likely something is to happen.

Connect

Suzanne puts 5 red tiles, 1 yellow tile, and 1 blue tile into a paper bag. Without looking, Marius draws a tile from the bag. What is the likelihood of each outcome?

- Marius draws a red tile.
- Marius draws a coloured tile.
- Marius draws a yellow tile.
- Marius draws a green tile.

Marius is more likely to draw a red tile than a blue tile.

There are 7 tiles in the bag.
Marius could pick any 1 of the 7 tiles.
There are 7 possible outcomes.

5 of the 7 tiles are red.
It is likely or **probable** that Marius draws a red tile.

All the tiles are coloured.
It is certain that Marius draws a coloured tile.
Marius always draws a coloured tile.

Only 1 of the 7 tiles is yellow.
It is **unlikely** or **improbable** that Marius draws a yellow tile.

There are no green tiles in the bag.
It is **impossible** for Marius to draw a green tile.
Marius will never draw a green tile.

Practice

1. Describe a situation that is:
   a) likely
   b) unlikely
   c) certain
   d) impossible
   e) probable
   f) improbable

2. a) Describe an event that always happens.
   b) Describe an event that never happens.

Unit 11 Lesson 1   **375**

3. Describe each outcome.
   Use these words: impossible, unlikely, likely, certain
   a) Someone in your class will win a raffle.
   b) Someone in your class is 10 years old.
   c) It will rain tomorrow.
   d) Your favourite team will win the Stanley Cup this year.
   e) You will have math homework next Wednesday.

4. There are 2 paper bags.
   Each bag has 1 red tile and 10 green tiles.
   Without looking, you draw 1 tile from each bag.
   Order these events from least likely to most likely.
   a) You draw 2 red tiles.
   b) You draw 2 coloured tiles.
   c) You draw 2 yellow tiles.
   d) You draw 2 green tiles.
   e) You draw 1 green tile and 1 red tile.

5. Alex and Rebecca spin the pointer on this spinner.
   Alex gets a point if the pointer lands on an even number.
   Rebecca gets a point if it lands on an odd number.
   Each person spins the pointer 20 times.
   The person with more points wins.
   Who is more likely to win? How do you know?

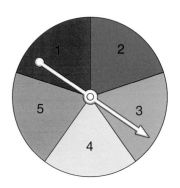

6. Design each spinner:
   a) Yellow is more likely than red.
      Red is more likely than blue.
   b) Blue and green are equally likely,
      but less likely than yellow.
   c) Yellow is certain.
   How did you decide on the number of sectors and the colours?

Reflect

What is the difference between an impossible event and an unlikely event? Use examples to explain.

At Home

Pick an event you think is likely. Ask family members if they think this event is certain, likely, unlikely, or impossible. Are their answers the same as yours? Explain.

LESSON

Conducting Experiments

 ···

Sum Fun

You will need 2 number cubes each labelled 1 to 6.

➤ Take turns to roll the number cubes.
➤ Find the sum of the 2 numbers rolled.
  If the sum is even, you score a point.
  If the sum is odd, your partner scores a point.
➤ Record the results in a table.
➤ The first player to score 20 points wins.
➤ Who do you think will have more points after 36 turns? Explain.

Show and Share

Compare your results with those of another pair of students. Explain any differences.
List the outcomes of the game.
Which is more likely, an even sum or an odd sum? How do you know?
How could you find the probability of an even sum? An odd sum?

Numbers Every Day

Number Strategies
Find each product:
0.71 × 100
7.1 × 100
71 × 100
710 × 100
What patterns do you see in the factors and the products?

LESSON FOCUS | Predict, then find, the probability of outcomes.   **377**

Connect

Jamie and Alexis are playing *Predicting Products*.
They take turns rolling 2 number cubes,
each labelled 1 to 6.
If the product of the 2 numbers rolled is odd,
Jamie gets a point.
If the product is even, Alexis gets a point.
The winner is the first person to get 20 points.
Who is more likely to win?

Jamie	Alexis
Odd Product	Even Product

Organize the possible outcomes in a table.
This will help you predict the winner.

From the table:
• There are 36 possible outcomes.
• 27 outcomes are even products.
• 9 outcomes are odd products.

×	1	2	3	4	5	6
1	1	2	3	4	5	6
2	2	4	6	8	10	12
3	3	6	9	12	15	18
4	4	8	12	16	20	24
5	5	10	15	20	25	30
6	6	12	18	24	30	36

You say: "The probability of getting an even product is 27 out of 36. The probability of getting an odd product is 9 out of 36."

You write:
Probability = $\frac{27}{36}$
Probability = $\frac{9}{36}$

The probability that Alexis wins is $\frac{27}{36}$.
The probability that Jamie wins is $\frac{9}{36}$.

Practice

1. Two number cubes are rolled. Both cubes are labelled 1 to 6.
   The numbers rolled are added.
   List the outcomes.
   What is the probability of each outcome?
   a) The sum is 12.
   b) The sum is less than 4.
   c) The sum is 7.
   d) The sum is 2.

2. Misha rolls two number cubes labelled 1 to 6.
   She finds the product of the 2 numbers rolled.
   Find the probability of each outcome.
   a) The product is 2.   b) The product is 36.
   c) The product is 12.   d) The product is 13.
   e) The product is greater than 25.

3. Dave and Bob play a game.
   Each tosses a coin.
   If the outcomes are the same, Dave scores a point.
   If the outcomes are different, Bob scores a point.
   The first person to score 20 points is the winner.
   a) List the outcomes.
   b) Who do you think will win? Explain.
   c) Do you think this is a fair game?
      Use pictures, words, or numbers to explain.

A **fair game** is one where all players are **equally likely** to win.

4. Vicki and Alastair have a spinner each.
   They play this game:
   • Each player spins her or his pointer.
   • If the pointers land on the same colour, Vicki scores a point.
   • If the pointers land on different colours, Alastair scores a point.
   a) Design spinners so Vicki is more likely to win.
   b) Design spinners so Alastair is more likely to win.
   c) Design spinners so the game is fair.

5. Design a fair game that uses 2 number cubes.
   Describe the rules of the game.
   How do you know the game is fair?

Reflect

What does it mean when we say,
"Two outcomes are equally likely."?
Use an example to explain.

Math Link

Your World

A meteorologist studies the weather.
She uses computers, satellites, and radar to predict if it will rain or snow.

ASSESSMENT FOCUS | Question 4

Unit 11 Lesson 2

LESSON 3

Probability and Fractions

This spinner has 6 equal sectors.
The probability the pointer
will land on blue is 3 out of 6, or $\frac{3}{6}$.
The probability it will land on yellow is $\frac{2}{6}$.
The probability it will land on red is $\frac{1}{6}$.

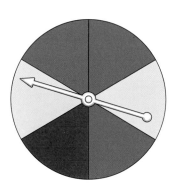

Explore

You will need 1-cm grid paper.

Design a square game board with 4 colours to meet these conditions.
- The probability of landing on a blue square is $\frac{5}{8}$.
- The probability of landing on a red square is $\frac{1}{8}$.
- The probability of landing on a yellow square is $\frac{1}{16}$.

Show and Share

Compare your game board with that of another pair of students.
How are they the same? How are they different?
What is the probability of landing on the fourth colour?
How could you make a game board with a different
number of squares?

Connect

Jenny and Maryann put coloured cubes into a bag.
They chose colours so:
- It is equally likely a red, green, or yellow cube is selected.
- The probability of drawing a blue cube is $\frac{2}{5}$.

How many cubes of each colour should Jenny and Maryann use?

The probability of drawing a blue cube is to be $\frac{2}{5}$.
So, 2 of every 5 cubes must be blue.
The total number of cubes must be a multiple of 5.
There are many ways to fill the bag.

➤ Suppose Jenny and Maryann used 5 cubes.
Two cubes must be blue.
Drawing a red, green, or yellow cube must be equally likely.
So, put 1 red, 1 green, and 1 yellow cube into the bag.

➤ Suppose Jenny and Maryann use 10 cubes.
$\frac{2}{5} = \frac{4}{10}$, so 4 of the 10 cubes must be blue.

The remaining 6 cubes must be divided equally among red, green, and yellow.
So, Jenny and Maryann placed 2 red, 2 green, and 2 yellow cubes in the bag.

Practice

1. A paper bag contains 2 green tiles, 4 yellow tiles, and 1 blue tile. Liz draws a tile without looking. What is the probability the tile is green? Yellow? Blue? Use fractions in your answers.

Numbers Every Day

Number Strategies

Find each sum:

12.87 + 0.15
128.7 + 0.15
1287 + 0.15

Unit 11 Lesson 3 **381**

2. Dave tossed a coin 20 times. Heads turned up 12 times.
   a) How many times did tails turn up?
   b) What fraction of tosses turned up heads?
   c) What fraction of tosses turned up tails?
   d) Are these results what you would expect? Explain.
   e) Dave tosses the coin 100 times.
      What would you expect the results to be? Explain.

3. Avril spins the pointer on this spinner several times.
   Here are her results.

   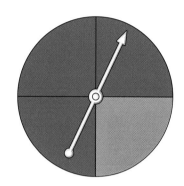

   a) How many times did Avril spin the pointer?
      How do you know?
   b) What fraction of the spins were blue?
   c) What fraction of the spins were orange?
   d) Were Avril's results what you would have expected? Explain.

4. A solid with 10 congruent faces is a decahedron.
   Shannon and Joshua roll a decahedron
   labelled 1 to 10.
   a) What is the probability Shannon rolls an odd number?
   b) Joshua says there is a probability of $\frac{1}{5}$ for rolling a number
      with a certain digit. What is the digit?

5. Sharma made this game board
   using foam Pattern Blocks.
   a) What is the probability of landing on yellow?
   b) Create a game board where the probability
      of landing on blue is $\frac{1}{4}$, and on yellow is $\frac{3}{4}$.

Reflect

How can we use fractions to describe probabilities?
Use pictures, numbers, and words to explain.

Probability in Games

Explore

What's the Difference?

You will need 2 number cubes, each labelled 1 to 6.

➤ Each player rolls 1 number cube.
➤ Find the difference of the numbers rolled.
   If the difference is less than 3,
   one player gets a point.
   If the difference is 3 or greater,
   the other player gets a point.
   The first player to score 10 points wins.
➤ Who do you think is more likely to win?
   Explain.
➤ Play the game 10 times. Record your results.
   How do your results compare to your prediction?
   Explain.

The difference is 3. That's one point for me!

Show and Share

Share your results with another pair of students.
Do you think this game is fair? Explain.
If not, how could you make it fair?
One of you is older than the other 3.
Does this affect the results? Explain.

Connect

Maude and Claude play a game with cards.
The standard deck of playing cards has 52 cards.
There are 4 suits: spades, diamonds, hearts, and clubs
Each suit has 13 cards.

LESSON FOCUS | Explore probability in games.

Maude picks a card and records the suit.
She replaces the card in the deck and shuffles the cards.
Claude picks a card.
If Claude's card is the same suit as Maude's,
he scores a point.
If Claude's card is a different suit,
Maude scores a point.
The player with the most points after 10 turns wins.

➤ Which player has a better chance of winning? Explain.

Suppose Maude draws a heart.
To score a point, Claude must draw a heart.
There are 13 hearts in the deck, and 39 other cards.
The probability Claude draws a heart is $\frac{13}{52}$.
The probability Claude does not draw a heart is $\frac{39}{52}$.
So, Maude has a better chance of scoring a point
than Claude.

➤ How could you change the game to make it fair?

Suppose Claude gets a point if his card
is the same colour as Maude's.
If the two cards are different colours,
Maude gets a point.

There are 26 red cards and 26 black cards
in the deck.
The probability Claude draws a red card is $\frac{26}{52}$.
The probability Claude draws a black card is $\frac{26}{52}$.
Now, Claude and Maude have equal chances
of scoring a point.

➤ Maude has played the fair game 10 times before.
Claude has never played this game before.
Does this affect the result?

When a game is fair, and all players have
equal chances of scoring, the fact that a person
has played many times before does not
affect the result.

Numbers Every Day

Number Strategies

Arrange the digits
2, 6, 9, and 0 so
the difference is
as close as possible
to 5.0.

☐.☐ − ☐.☐

Practice

1. Brandon and Carla play this game:
   Each rolls a number cube labelled 1 to 6.
   They use the numbers rolled to make a fraction.
   Brandon's number is the numerator.
   Carla's number is the denominator.
   Carla wins a point if the fraction is less than or equal to 1.
   Brandon wins a point if the fraction is greater than 1.
   Who do you think has the better chance of winning? Explain.
   Play the game with a partner.
   Did your results agree with your prediction?

2. Design a fair game that involves two people tossing 3 coins.
   a) Describe the rules of your game.
   b) Explain how you know the game is fair.
   c) Suppose Jake and Jennifer play this game.
      Jennifer is 10 years older than Jake.
      Jennifer wins.
      Did she win because she is older? Explain.

3. The *Game of Pig* involves strategy and probability.
   Here are the rules:
   • Roll two number cubes each labelled 1 to 6.
     Find the sum of the numbers rolled.
     These are the points you score.
   • You may choose to stop, or you may roll again.
     If you roll a double, your turn is over.
     You score no points for your turn.
   • The first player to score 100 points wins.
   What strategies could you use to win *The Game of Pig*?

Reflect

Describe a game you have played that involves strategy and probability.

List 5 games you like to play.
Which games involve probability?
What other factors affect who wins the game?

Assessment Focus | Question 2

LESSON 5

Strategies Toolkit

Explore

Ali, Brian, and Caitlin are to be photographed together.
How many different ways can they be arranged in a line?
What is the probability that Brian and Caitlin will be next to each other?

Show and Share

Describe the strategy you used to solve this problem.

Connect

Mayhew School has 4 championship banners to hang in a hallway.
The banners for basketball, volleyball, cross-country, and track and field are hung in line.
How many different ways can the banners be hung?
What is the probability that the banner for basketball will be next to that for volleyball?

Strategies

- Make a table.
- Use a model.
- Draw a diagram.
- Solve a simpler problem.
- Work backward.
- Guess and check.
- Make an organized list.
- Use a pattern.
- Draw a graph.

What do you know?
- There are 4 different banners.
- The banners hang in line.

Think of a strategy to help you solve the problem.
- You can **use a model**.

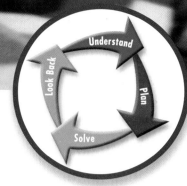

Use a tree diagram as a model.
Record the different arrangements.
The tree diagram is started below.

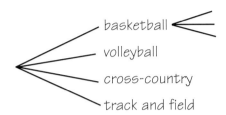

Copy and complete the diagram.
How many different arrangements are there?
What is the probability that the banners for basketball and volleyball are beside each other?

How could you have solved this problem another way?

Practice

Choose one of the **Strategies**

1. Matthias is framing a photo of himself to give to his mother as a present. He will use:
   - a photo of himself at home, at school, or playing baseball
   - a white border or a black border
   - a silver frame or a wood frame

   How many different presents can he make?

2. Mr. Roe has cards labelled 1, 2, and 3.
   He arranges the cards to make a 3-digit number.
   What is the probability that the 3-digit number is less than 200?

Reflect

How can using a model help you solve a problem?
Use an example to explain.

Unit 11 Lesson 5

Unit 11   Show What You Know

LESSON

1 2 3

1. Each letter of the word MATHEMATICS is written on a card.
   The cards are shuffled. One card is drawn without looking.
   a) List the outcomes.
   b) Which letter is most likely to be drawn?
   c) Which letter is least likely to be drawn?
   d) What is the probability of drawing each letter?

1 2

2. Order these events from most likely to least likely.
   State each probability as a fraction.
   a) The pointer on this spinner landing on red
   b) Tossing a coin and getting heads
   c) Rolling a number cube and getting 5
   d) Drawing a diamond from a standard
      deck of playing cards

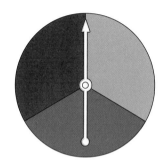

2

3. Two number cubes are rolled. Each is labelled 1 to 6.
   The numbers on the cubes are multiplied.
   List the outcomes.
   Find the probability of each outcome.
   a) The product of the numbers rolled is 1.
   b) The product is 30.
   c) The product is less than 6.
   d) The product is greater than 7.

2 3 4

4. Lynda has a standard deck
   of 52 playing cards.
   She shuffles the deck, then draws a card
   without looking.
   a) What is the probability Lynda draws a red card?
   b) What is the probability Lynda draws a 4?
   c) What is the probability Lynda draws
      the 4 of diamonds?

5. Troy rolled a number cube 36 times.
A 6 was rolled 11 times.
1, 2, 3, 4, and 5 were rolled the same number of times.
a) What fraction of rolls were a 3?
b) What fraction of rolls were an even number?
c) Are these results different from what you would expect? Explain.

6. In the game *Monopoly*, Ali rolls 2 number cubes, each labelled 1 to 6.
a) Ali must roll a double to get out of jail. What is the probability he will roll a double? How do you know?
b) Ali has played *Monopoly* 100 times. Does this affect the probability that he will roll a double? Explain.
c) Ali's mother has played *Monopoly* 1000 times. Does her age affect the probability that she will roll a double to get out of jail? Explain.

UNIT 11 Learning Goals

- ✓ use the language of probability
- ✓ list the outcomes of an experiment
- ✓ conduct experiments, predict results, and explain results
- ✓ use fractions to describe probability
- ✓ show that results are not affected by age, experience, or skill

Unit Problem

At the Pet Store!

Work with a partner.
You will need:
- Snap Cubes
- a paper bag
- Bristol board

Part 1 Lucky Silver

A large tank contains 50 fish.
Suppose you pick a fish at random.
- The probability you pick a silver fish is $\frac{1}{10}$.
- The probability you pick a red fish is $\frac{2}{5}$.
- It is equally likely that you will pick one of the remaining orange, blue, black, white, or pink fish.

How many fish of each colour are in the tank?

Part 2 Red, Yellow, and Blue

➤ Choose 3 colours of Snap Cubes to represent 3 different colours of fish. Put cubes into a bag to match these probabilities:
  - The probabilities of picking a yellow fish and a blue fish are the same.
  - The probability of picking a red fish is greater.
➤ Try your experiment. Without looking, draw a cube from the bag. Have your partner record the colour. Return the cube to the bag.

➤ Draw a cube 10 times.
  Switch roles.
  Do your results match the stated probabilities?
  Explain.

Part 3

Suppose you are starting a fish tank.
➤ Choose 3 or 4 kinds of fish.
  • Decide on the probability of picking each kind of fish.
  • Decide how many fish you will have altogether.
    How many fish of each kind will there be?
➤ Make a poster of your fish tank on the top half of the Bristol board.
  On the bottom half, explain:
  • how you chose the kind of fish
  • how you decided on the probabilities
  • how you found the number of fish of each kind

Check List

Your work should show
☑ specific answers to each of the probability questions, using fractions where appropriate
☑ any tables and diagrams you use to find and record your answers
☑ clear explanations of your procedures and results
☑ correct use of the language of probability

Reflect on the Unit

How can you use probability to predict outcomes?
Use pictures, words, and numbers to explain your answer.

Cross Strand Investigation

The Domino Effect

You will need dominoes, a metre stick, a stopwatch, and grid paper.

Part 1

➤ Begin with 20 dominoes.
  Stand them on end, 3 cm apart.
  Use a stopwatch.
  Push one domino at one end, so all the dominoes fall.
  Time how long it takes them to fall.
  Record the number of dominoes and the time in a table.

➤ Repeat with 30 dominoes, 40 dominoes, 50 dominoes, up to 80 dominoes.

➤ Describe any patterns you see in the table.

➤ Predict how long it would take 120 dominoes to fall.
  How did you make your prediction?

Part 2

Draw a broken-line graph to display the data in your table.
Describe the graph.
About how long would it take 35 dominoes to topple?
How do you know?

Display Your Work

Report your findings using pictures, numbers, and words.

Take It Further

Investigate different arrangements of dominoes.
What effect does placing the dominoes closer together
have on the time it takes them to topple? Explain.
Arrange the dominoes in a curve.
How long does it take them to topple?

Units 1–11 Cumulative Review

UNIT

1

1. Here is the table for an Input/Output machine.
   a) The patterns continue.
      Write the next 3 input and output numbers.
   b) Draw the Input/Output machine for this table.

Input	Output
99	33
96	32
93	31
90	30

2

2. Find each result.
   a) 2206   b) 3436   c) 3689
      + 985   − 3351   − 468

   d) 73 × 24   e) 7808 ÷ 8   f) 44 × 82

3. Ryan earned $162 babysitting and doing odd jobs.
   He earned $5 per hour for 12 h of babysitting.
   How much did he earn doing odd jobs?

3

4. Is a rhombus a regular polygon?
   How do you know?

5. Name each triangle.
   How did you choose each name?
   a)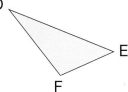
   b)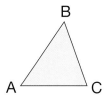
   c)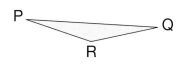

4

6. Add, subtract, multiply, or divide.
   a) 4.68 − 3.2   b) 48.7 ÷ 10   c) 153.9 ÷ 10   d) 12.4 − 3.03
   e) 4.68 × 100   f) 5.8 + 2.02   g) 0.56 + 5.6   h) 20.31 × 10

7. A decimal with hundredths rounds to 2.
   What might the decimal be? Give 3 different answers.

394

UNIT 5

8. The table shows the number of Canadians who visited various countries in 2002. Display these data in a graph.
Which type of graph did you choose to draw? Why?

Country	Canadian Visitors (thousands)
Australia	108
China	140
Cuba	331
France	505
Germany	255
Mexico	607
United Kingdom	720

9. Use SI notation to write:

   a) the exact time
   b) the time to the nearest minute

10. This object was made with centimetre cubes. Find the volume of the object in cubic centimetres and in millilitres.

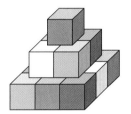

11. Monique bought groceries for $71.29. She paid with two $20 bills and four $10 bills. She got this change: a $5 bill, a toonie, a loonie, 2 quarters, 3 dimes, and a penny
Did Monique get the correct change? How do you know?

12. Copy this figure on 1-cm grid paper. Draw the image after each transformation.
   a) a reflection in the broken line
   b) a $\frac{1}{2}$ turn clockwise about the dot
   c) a translation of 6 squares left

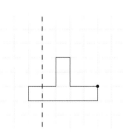

395

UNIT 7

13. Use dot paper.
  a) Draw a polygon on dot paper. Label it A.
  b) Draw a polygon that is congruent to Polygon A. Label it B.
  c) For which transformation is Polygon A an image of Polygon B?
  d) Continue to draw congruent polygons. Does the polygon tessellate? Explain.

14. Copy the figure and the broken line on dot paper.
Use the broken line as a mirror line.
Draw the mirror image of the figure so the two figures form a new figure.
Is the new figure symmetrical?
How do you know?

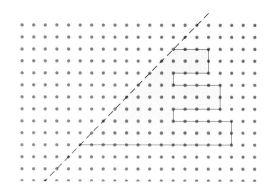

UNIT 8

15. Copy each statement.
Write a fraction to make each statement true.
  a) $1\frac{4}{5} < \square$
  b) $\frac{7}{10} < \square$
  c) $\square < \frac{6}{5}$
  d) $\frac{13}{4} < \square$

16. Write each fraction or mixed number as a decimal.
Then write the decimals in order from least to greatest.
$\frac{15}{4}, \frac{4}{5}, \frac{3}{10}, \frac{9}{2}, 2\frac{1}{10}$

17. Draw a picture to represent each decimal.
Then write 2 fractions for each decimal.
  a) 0.4
  b) 0.35
  c) 0.9
  d) 0.25

18. Find each product or quotient.
  a) 8.63×2
  b) $29.82 \div 7$
  c) $45.4 \div 5$
  d) 3.02×9

19. An ice cream cone costs $1.97.
Hunter bought 5 ice cream cones.
How much did Hunter pay?

UNIT 9

20. Find the area and the perimeter of each figure.
Write each perimeter in a different unit.

a)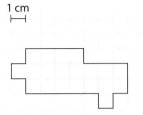

b)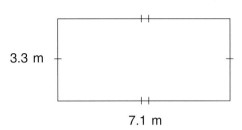

21. What might the dimensions of each rectangle be?
Give 3 different answers.
   a) a rectangle with perimeter 42 cm
   b) a rectangle with area 36 m²

UNIT 10

22. Multiply 30 × 4.
Use the product 30 × 4 to find each missing factor.
   a) 30 × □ = 150
   b) □ × 30 = 240
   c) 900 = 30 × □
   d) 30 × □ = 330

23. Use dot paper or grid paper.
Draw two figures that you think together will make a tiling pattern.
Cover the paper with the tiling pattern.
Colour the pattern.
Describe the smallest part of the pattern that repeats.

UNIT 11

24. Keltie rolls two number cubes.
Each number cube is labelled 3 to 8.
The numbers are multipled.
List the outcomes.
Find the probability of each outcome.
   a) The product of the numbers rolled is 40.
   b) The sum of the numbers rolled is less than 8.
   c) The product of the numbers rolled is 65.

397

Illustrated Glossary

a.m.: A time between midnight and just before noon.

Angle: Two straight lines meet to form an angle. Each side of an angle is called an *arm*. We show an angle by drawing an arc.

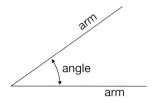

Area: The amount of surface a figure or region covers. We measure area in square units, such as square centimetres or square metres.

Arm: See Angle.

Axis (plural: axes): A number line along the edge of a graph. We label each axis of a graph to tell what data it displays.

Bar graph: Displays data by using bars of equal width on a grid. The bars may be vertical or horizontal.

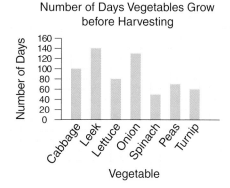

Base: The face that names a solid. For example, in this triangular prism, the bases are triangles.

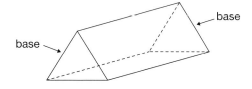

Bias: A graph with bias shows data in a way that someone else wants you to see them. A sample with bias does not truly represent the group.

Broken-line graph: A graph showing data points joined by line segments.

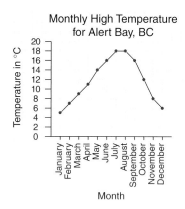

Capacity: A measure of how much a container holds. We measure capacity in litres or millilitres.

Centimetre: A unit used to measure length.
We write one centimetre as 1 cm.
1 cm = 0.01 m

Changing-step growing pattern: A number pattern where the number added increases.

Circumference: The distance around a circular object.

Clockwise: The hands on a clock turn in a clockwise direction.

Clockwise

Combination: A selection of items from different groups to make a smaller group.

Compatible numbers: Pairs of numbers you can easily divide mentally.

Compensation: Adding an amount to one term that you subtract from another.
27 + 53 = (27 + 3) + (53 − 3)
= 30 + 50
= 80

Composite number: A number that has more than two factors.

Congruent figures: Two figures that have the same size and shape.

Coordinates: Describe a location on a grid, using numbers to label vertical and horizontal grid lines. The coordinates (5,9) show the location of one point of this star.

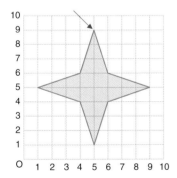

Core: See Repeating pattern.

Counterclockwise: A turn in the opposite direction to the direction the hands on a clock turn.

Counterclockwise

Cube: A solid with 6 faces that are congruent squares. Two faces meet at an edge.

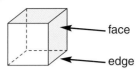

Cubic centimetre (cm³): A unit to measure volume and capacity. A centimetre cube has a volume of one cubic centimetre.
We write one cubic centimetre as 1 cm³.
1 cm³ = 1 mL

Data: Information collected from a survey or experiment.

Decagon: A polygon with 10 sides.

Decahedron: A 10-sided solid.

Decimal: A way to write a fraction or mixed number. The mixed number $3\frac{2}{10}$ can be written as the decimal 3.2.

Decimal point: Separates the whole number part and the fraction part in a decimal. We read the decimal point as "and." We say 3.2 as "three **and** two-tenths."

Decimetre: A unit to measure length. We write one decimetre as 1 dm.
1 dm = 0.1 m and 1 dm = 10 cm

Degree: A unit to measure temperature. We write one degree Celsius as 1°C.

Denominator: The part of a fraction that tells how many equal parts are in one whole. The denominator is the bottom number in a fraction.

Diagonal: A line that joins opposite corners or vertices of a figure.

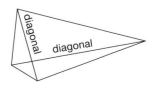

Difference: The result of a subtraction. The difference of 5 and 2 is 3. 5 − 2 = 3.

Dimensions:
1. The measurements of a figure or object. A rectangle has 2 dimensions, length and width. A cube has 3 dimensions, length, width, and height.
2. For an array, the dimensions tell the number of rows and the number of columns.

Displacement: The volume of water moved or displaced by an object put in the water. The displacement of this cube is 50 mL.

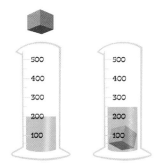

Dividend: The number to be divided. In the division sentence 77 ÷ 11 = 7, the dividend is 77.

Divisor: The number by which another number is divided. In the division sentence 77 ÷ 11 = 7, the divisor is 11.

Dodecahedron: A 12-sided solid.

Edge: Two faces of a solid meet at an edge. See also Cube, Prism, and Pyramid.

Elapsed time: The amount of time that passes from the start to the end of an event. The elapsed time between when you eat lunch and the end of school is about 3 h.

Equally likely: The outcomes of an event that are equally probable. For example, if you toss a coin, it is equally likely that the coin will land heads up as tails up.

Equally probable: See Equally likely.

Equation: Uses the = symbol to show two things that represent the same amount. 5 + 2 = 7 is an equation.

Equilateral triangle: A triangle with all sides equal.

Equivalent decimals: Decimals that name the same amount. 1.4 and 1.40 are equivalent decimals.

Estimate: Close to an amount or value, but not exact.

Experiment: In probability, a test or trial used to investigate an idea.

Face: One side of a solid. See also Cube, Prism, and Pyramid.

Factor: Numbers that are multiplied to get a product. In the multiplication sentence 3 × 7 = 21, the factors of 21 are 3 and 7.

Fair game: A game where all players have the same chance of winning.

Fibonacci number: Any number in the Fibonacci sequence.

Fibonacci sequence: 1, 1, 2, 3, 5, 8, 13, …. Each number after the first pair is the sum of the two numbers before it.

Formula: A short way to state a rule. Area = length × width is a formula for the area of a rectangle.

Gram: A unit to measure mass. We write one gram as 1 g. 1000 g = 1 kg

Grid: See Coordinates.

Growing pattern: A pattern where each term or frame is greater than the previous term or frame.

Frame 1   Frame 2   Frame 3

1, 3, 8, 10, 15, 17, 23, …

Hexagon: A polygon with 6 sides.

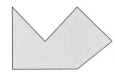

Hundredth: A fraction that is one part of a whole when it is divided into 100 equal parts. We write one-hundredth as $\frac{1}{100}$ or 0.01.

Image: The figure that is the result of a transformation. This is a rectangle and its image after a translation of 6 right and 1 up.

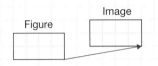

Improbable: An event that is unlikely to happen but not impossible.

Icosahedron: A 20-sided solid.

Improper fraction: A fraction that shows an amount greater than one whole. The numerator is greater than the denominator. $\frac{3}{2}$ is an improper fraction.

Input/Output machine: Performs an operation on a number (the input) to produce another number (the output). This Input/Output machine divides the input by 2.

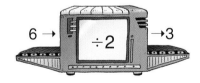

Intervals: Groups of data sorted to be more manageable. When looking at a large set of data, intervals have equal possible ranges. The interval 50–59 has the same range as the interval 60–69

Irregular polygon: A polygon that does not have all sides equal or all angles equal. Here are two irregular hexagons.

Isosceles triangle: A triangle with two sides equal.

Key: See Pictograph.

Kilogram: A unit to measure mass. We write one kilogram as 1 kg.
1 kg = 1000 g

Kilometre: A unit to measure long distances. We write one kilometre as 1 km. 1 km = 1000 m

Kite: A quadrilateral where two pairs of adjacent sides are equal.

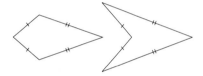

Line of symmetry: Divides a figure into two congruent parts. If we fold the figure along its line of symmetry, the parts match.

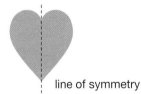

line of symmetry

Linear dimension: Length, width, depth, height, thickness.

Litre: A unit to measure the capacity of a container. We write one litre as 1 L.
1 L = 1000 mL

Mass: Measures how much matter is in an object. We measure mass in grams or kilograms.

Metre: A unit to measure length. We write one metre as 1 m.
1 m = 100 cm

Milligram: A unit to measure mass. We write one milligram as 1 mg. 1000 mg = 1 kg

Millilitre: A unit to measure the capacity of a container. We write one millilitre as 1 mL. 1000 mL = 1 L. 1mL = 1 cm³

Millimetre: A unit to measure length. We write one millimetre as 1 mm. One millimetre is one-tenth of a centimetre; 1 mm = 0.1 cm. 10 mm = 1 cm

Mirror line: A line through which a figure is reflected. See Reflection.

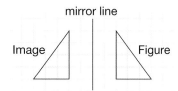

Mixed number: Has a whole number part and a fraction part. $3\frac{1}{2}$ is a mixed number.

Multiple: Start at a number, then count on by that number to get the multiples of that number. To get the multiples of 3, start at 3 and count on by 3: 3, 6, 9, 12, 15, …

Multiplication fact: A sentence that relates factors to a product. $3 \times 7 = 21$ is a multiplication fact.

Net: An arrangement that shows all the faces of a solid, joined in one piece. It can be folded to form the solid.

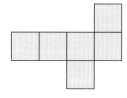

Non-standard units: Floor tiles, car lengths, and strides are some non-standard units that could be used for measuring length.

Number line: Has numbers in order from least to greatest. The spaces between pairs of consecutive numbers are equal.

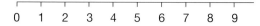

Numerator: The part of a fraction that tells how many equal parts to count. The numerator is the top number in a fraction. In the fraction $\frac{2}{3}$, the numerator is 2. We count 2 thirds of the whole.

Octagon: A polygon with 8 sides.

Operation: Something done to a number or quantity. Addition, subtraction, multiplication, and division are operations.

Outcome: One result of an event or experiment. Tossing a coin has two possible outcomes, heads or tails.

p.m.: A time between noon and just before midnight.

Parallel: Two lines that are always the same distance apart are parallel.

Parallelogram: A quadrilateral, where 2 pairs of opposite sides are parallel.

Pattern rule: Describes how to make a pattern. For the pattern 1, 2, 4, 8, 16, …, the pattern rule is: Start at 1. Multiply by 2 each time.

Perimeter: The distance around a figure. It is the sum of the side lengths. The perimeter of this rectangle is: 2 cm + 4 cm + 2 cm + 4 cm = 12 cm.

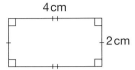

Perpendicular: Two lines that intersect at a right angle are perpendicular.

Pictograph: Uses pictures and symbols to display data. Each picture or symbol can represent more than one object. A key tells what each picture represents.

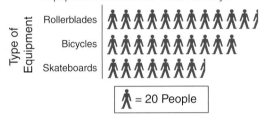

Plane of symmetry: The cut that is made through a solid, so one part of the solid is the mirror image of the other.

Placeholder: A zero used to hold the place value of the digits in a number. For example, the number 603 has 0 tens. The digit 0 is a placeholder.

Polygon: A closed figure with three or more straight sides. We name a polygon by the number of its sides. For example, a five-sided polygon is a pentagon.

Prediction: You make a prediction when you decide how likely or unlikely it is that an event will happen.

Prime number: A number that has exactly two factors.

Prism: A solid with 2 bases.

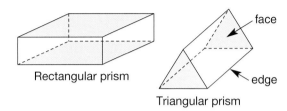

Probability: Tells how likely it is that an event will occur. Rolling a number cube labelled 1 to 6 has 6 equally probable results. The probability that you will roll a 2 is $\frac{1}{6}$.

Probable: An event that is likely but not certain to happen.

Product: The result of a multiplication. The product of 5 and 2 is 10. $5 \times 2 = 10$.

Proper fraction: Describes an amount less than one. A proper fraction has a numerator that is less than its denominator. $\frac{5}{7}$ is a proper fraction.

Pyramid: A solid with 1 base.

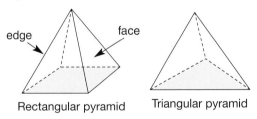

Quotient: The number obtained by dividing one number into another. In the division sentence $77 \div 11 = 7$, the quotient is 7.

Quadrilateral: A figure with 4 sides.

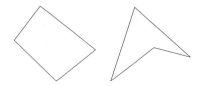

403

Range: Tells how spread out the numbers in a set of data are. We find the range by subtracting the least value from the greatest value. The range in the children's heights is 132 cm – 110 cm = 22 cm.

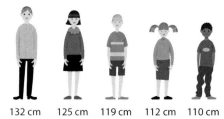

Rectangle: A quadrilateral, where 2 pairs of opposite sides are equal and each angle is a right angle.

Rectangular prism: See Prism.

Rectangular pyramid: See Pyramid.

Reflection: Reflects a figure in a mirror line to create a congruent image. This is a triangle and its reflection image.

Regular figure: See Regular polygon.

Regular polygon: A regular polygon has all sides equal and all angles equal. Here is a regular hexagon.

Related facts: Sets of addition and subtraction facts or multiplication and division facts that have the same numbers. Here are two sets of related facts:

2 + 3 = 5   5 × 6 = 30
3 + 2 = 5   6 × 5 = 30
5 − 3 = 2   30 ÷ 6 = 5
5 − 2 = 3   30 ÷ 5 = 6

Remainder: What is left over when one number does not divide exactly into another number. For example, in the quotient 13 ÷ 5 = 2 R3, the remainder is 3.

Repeating pattern: A pattern with a core that repeats. The core is the smallest part of the pattern that repeats. In the pattern: 1, 8, 2, 1, 8, 2, 1, 8, 2, …, the core is 1, 8, 2.

Rhombus: A quadrilateral, where all sides are equal and 2 pairs of opposite sides are parallel.

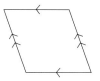

Right angle: Two lines that meet in a square corner make a right angle.

Rep-tile: A polygon that can be copied and arranged to form a larger, similar polygon.

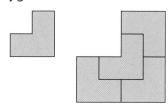

Rotation: Turns a figure about a turn centre. This is a triangle and its image after a rotation of a $\frac{1}{4}$ turn about one vertex:

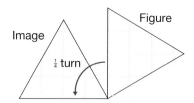

Round: To estimate a number to a certain place value. For example, 397 482 rounded to the nearest thousand is 397 000.

Same-step growing pattern: A number pattern where the same number is added each time.

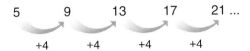

Sample: A small group chosen from the entire group. Samples are often used for surveys.

Scale: 1. The numbers on the axis of a graph show the scale.
2. The drawing of an object is to scale if the drawing and the object are similar. The drawing is larger or smaller than the object but has the same shape.

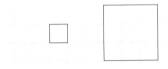

Scalene triangle: A triangle with no sides equal.

Second: A small unit of time. There are 60 seconds in 1 minute. 60 s = 1 min

Sector: A portion of a circle that extends from the centre to the edge.

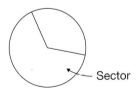

SI (International System of Units) notation: A standard way to give values for time, date, mass, volume, and other quantities.

Similar figures: Two figures are similar when they have the same shape. Similar figures do not have to be the same size. These hexagons are similar.

Solid: Has length, width, and height. Solids have faces, edges, vertices, and bases. We name some solids by the number and shape of their bases.

Pentagonal pyramid   Hexagonal prism

Square: A quadrilateral with equal sides and each angle is a right angle.

Speed: A measure of how fast an object is moving.

Square centimetre: A unit of area that is a square with 1-cm sides. We write one square centimetre as 1 cm².

Square metre: A unit of area that is a square with 1-m sides. We write one square metre as 1 m².

Standard units: Metres, square metres, cubic metres, kilograms, and seconds are some standard units.

Sum: The result of addition.
The sum of 5 and 2 is 7.
5 + 2 = 7.

Survey: Used to collect data. You can survey your classmates by asking them which is their favourite ice cream flavour.

405

Symmetrical: A figure is symmetrical if it has one or more lines of symmetry.

Tenth: A fraction that is one part of a whole when it is divided into 10 equal parts. We write one-tenth as $\frac{1}{10}$ or as 0.1.

Term: One number in a number pattern. For example, the number 4 is the third term in the pattern 1, 2, 4, 8, 16, ….

Tessellation: A tiling pattern with all figures congruent.

Tiling pattern: A pattern of tiles that covers a surface with no gaps or overlaps.

Tonne: A unit used to measure a very large mass. We write one tonne as 1 t. 1 t = 1000 kg

Transformation: A translation (slide), a reflection (flip), and a rotation (turn) are transformations.

Translation: Slides a figure from one location to another. A translation arrow joins matching points on the figure and its image. This figure has been translated 6 squares left and 2 squares up.

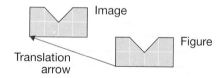

Translation arrow: See Translation.

Trapezoid: A quadrilateral with 1 pair of sides parallel.

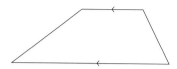

Tree diagram: A visual representation of all outcomes when there is more than one choice to be made. Flipping a coin twice leads to 4 possible outcomes.

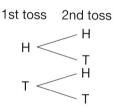

Triangular prism: See Prism.

Triangular pyramid: See Pyramid.

Turn centre: See Rotation.

Vertex (plural: vertices):
1. A point where two sides of a figure meet.
2. A point where two or more edges of a solid meet.

Volume: The amount of space occupied by an object. Volume can be measured in cubic centimetres (mL). 1 cm³ = 1 mL

Index

Numbers and Symbols
3- and 4-digit numbers
 adding, 37, 38
4-digit numbers
 subtracting, 43, 44
24-hour clock, 193, 194

A
Adding
 3- and 4-digit numbers, 37, 38
 decimals, 128, 129
 expanded form, 38
 with mental math, 34, 35
Angles, 80
 in regular polygons, 81
AppleWorks
 creating spreadsheets, 162, 163
 drawing graphs, 167, 168, 173, 174
 exploring tiling patterns, 365–367
Area, 321, 322
 estimating, 337, 338
 of irregular polygon, 333–335
 of rectangle, 329, 330

B
Bar graph, 155
 drawing, 164, 165
 drawing with *AppleWorks*, 167, 168
Base Ten Blocks, 28
 adding decimals, 128, 129
 dividing decimals, 141, 142
 multiplying 2-digit numbers, 60
 multiplying decimals, 138
 subtracting decimals, 133, 134
Bridges, 78, 106
Broken-line graph, 169-171
 drawing with *AppleWorks*, 173, 174

C
Calculator
 writing fractions as decimals, 282
Capacity, 208
 and volume, 213, 214
Centimetres (cm), 314
Certain event, 374, 375
Circumference, 319
Clockwise, 233
Comparing numbers, 28, 29
Compatible numbers, 57
Compensation, 35
Composite numbers, 31, 32
Congruent figures, 91
 tessellations of, 244
Coordinate grids, 248, 249
Coordinates, 248, 249
Counterclockwise, 233
Cubic centimetre (cm^3), 211, 214

D
Data
 interpreting, 154–156
Decimal benchmarks, 276, 277
Decimals, 112, 113
 adding, 128, 129
 comparing and ordering, 118–120
 dividing by 10, 141–143
 dividing with hundredths, 299, 300
 dividing with tenths, 296, 297
 multiplying by 10 and 100, 137–139
 multiplying with hundredths, 291–292
 multiplying with tenths, 287–288
 relation to fractions, 272–274
 rounding, 122, 123, 125, 126
 subtracting, 133, 134
Decimetres (dm), 314
Decreasing line graph, 171

Differences
 estimating, 125, 126
Displacement, 214
Distance, 197, 198
 around circular objects, 318, 319
Dividend, 47, 284
Dividing, 68, 69
 decimals by 10, 141–143
 with mental math, 58
 with whole numbers, 64–66
Division
 relating to fractions, 279, 280
 short, 66
Division facts, 46, 47
Divisor, 47, 284

E
Equilateral triangle, 84, 85
Equivalent decimals, 116
Equivalent fractions, 260–262
Escher, M.C., 226
Expanded form
 of numbers, 28, 29
 using to add, 38

F
Factor, 32, 46, 51, 348–350
Fair, 176
 game, 379, 383
Fibonacci (*also* Leonardo of Pisa), 356
Fibonacci number, 357
Fibonacci sequence, 356
Flip (*see* Reflection)
Fraction benchmarks, 276, 277
Fractions, 264, 265
 and probability, 380, 381
 comparing and ordering, 267, 268
 equivalent, 260–262
 improper, 264
 proper, 264
 relating to decimals, 272–274
 relating to divisions, 279–280

407

Frequency, 159, 165
  table, 159, 165
Front-end estimation, 34

G
Games, 396
  Auction Game, The, 205
  Fractions In-Between, 283
  Less Is More, 71
  Make 2!, 131
  Multiplication Tic-Tac-Toe, 50
  Order Up!, 271
  Spinning Decimals, 132
  Target No Remainder!, 67
  What's the Difference?, 383
  Who Has the Greater Product?, 56
Grams (g), 217
Graphs
  bar, 155, 164, 165
  broken-line, 169–171
  drawing with *AppleWorks*, 167, 168, 173, 174
  pictograph, 155

H
Height, 313
Hexagon, 81
  area of, 334, 335
  tessellations of, 244
Hundredths, 113

I
Image
  reflection, 237
  rotation, 233
  translation, 229
Impossible event, 375
Improbable event, 375
Improper fraction, 264
Increasing broken-line graph, 171
Inferences, 178
Input/Output machine, 9, 10, 20, 21
Intervals, 159, 164
Irregular polygon
  area of, 333-335
Isosceles triangle, 84, 85

K
Key, 155
Kilograms (kg), 217, 219, 220
Kilometres, 314

L
Length
  estimating with non-standard units, 316, 317
Leonardo of Pisa (*see* Fibonacci)
Likelihood of events, 374, 375
Line plot, 159
Line symmetry (*also* mirror line), 240, 241
  in regular polygons, 80–82
Linear dimension, 310, 311
Litre (L), 208

M
Mass, 216, 217, 219, 220
Math Link, 10, 42, 81, 90, 136, 176, 216, 315, 332, 358, 383
Mental math
  adding with, 34, 35
  dividing with, 58
  dividing decimals, 143
  multiplying with, 54, 55
  multiplying decimals, 139
  subtracting with, 40, 41
Metres (m), 314
Milligrams (mg), 217
Millilitres (mL), 208, 213, 214
Millimetres (mm), 314
Minutes (min), 191
Mirror line (*see* line symmetry)
Mixed number, 112, 113, 264, 265
Modelling patterns, 12, 13
Money
  estimating and counting, 202, 203, 205, 206
Multiples of 10
  multiplying with, 51, 52
Multiplication
  patterns in, 348–350
Multiplication facts, 46, 47
Multiplying, 68, 69
  decimals by 10 and 100, 137–139
  with mental math, 54, 55
  with multiples of 10, 51, 52

N
Non-standard units, 316, 317
Number patterns, 6, 7, 9, 356, 357
Numbers
  3- and 4-digits, 37, 38
  comparing, ordering, and representing, 28, 29
  composite, 31, 32
  prime, 31, 32

O
Octagon, 81, 88
Ones, 113
Operations, 9, 20, 21
Ordered pair, 249
Ordering numbers, 28, 29
Outcome, 374, 375

P
Parallelogram, 93
Pattern rule, 6, 7, 9, 13, 16, 17
Patterns
  graphing, 352, 353
  in multiplication, 348–350
  modelling, 12, 13
  solving problems with, 16, 17
  tiling, 243, 244, 362, 363
Pedometer, 43
Pentagons, 81, 88
  in tiling patterns, 363
Perimeter
  finding with grid paper, 322
  of polygons, 325, 326
  of rectangle, 329-330
Pictograph, 155
Placeholder, 143

Place-value chart, 28, 29
    adding decimals, 129
    dividing decimals, 142
    multiplying, 52
    multiplying decimals, 138
    showing decimals, 113
    subtracting decimals, 134
Plane, 100
Plane of symmetry, 100
Polygons
    identifying and naming, 80, 81
    perimeter of, 326
    regular, 81, 82, 333
Population, 175
Prime numbers, 31, 32
Probability, 371, 378
    and fractions, 380, 381
    in games, 383, 384
Probable event, 375
Product, 46, 51, 52, 54, 55, 348–350
Proper fraction, 264

Q
Quadrilaterals, 80, 81
Quotient, 47, 284
    estimating, 58

R
Range, 154, 155
Rectangle, 92
    area of, 329, 330
    perimeter of, 329, 330
Reflection, 236, 237
Regular polygon, 81, 82, 333

Representing numbers, 28, 29
Rep-tile, 256
Right angle, 87
Rotation, 232–234
Rounding, 34, 40
    decimals, 122, 123, 125, 126

S
Sample, 175
Scale, 155, 326
Scalene triangle, 84, 85
Seconds (s), 190
Short division, 66
SI notation, 191
Similar figures, 13, 256
Slide (*see* Translation)
Speed, 198
Spreadsheets
    creating with *AppleWorks*, 162, 163
Square, 82
Square pyramid, 97, 100
Standard form of numbers, 28, 29
Standard units, 317
Subtracting
    decimals, 133, 134
    with mental math, 40, 41
Sums
    estimating, 125, 126
Surveys
    interpreting results of, 175, 176
Symmetrical, 241

T
Tables, 154, 156, 178
Tally, 158, 159

Tenths, 115
Tessellation, 241, 254, 255
Tiling pattern, 243, 244, 362, 363
    exploring with *AppleWorks*, 365–367
Time
    measuring, 190, 191
    writing, 193, 194
Tonne (t), 219, 220
Transformations, 237
Translation, 228, 229
Translation arrow, 229
Trapezoid, 87
Triangles, 84
    constructing, 84
Triangular prism, 96, 103
Triangular pyramid, 95
Truss patterns, 78, 79, 106, 107
Turn (*see* Rotation)
Turn centre, 233

U
Unlikely event, 375

V
Vertex, 80
Volume, 210, 211
    and capacity, 213, 214

W
Whole numbers
    dividing, 64–66
    multiplying, 60–62
Word form of numbers, 28, 29

Acknowledgments

The publisher wishes to thank the following sources for photographs, illustrations, and other materials used in this book. Care has been taken to determine and locate ownership of copyright material in this text. We will gladly receive information enabling us to rectify any errors or omissions in credits.

Photography

Cover: Cornelia Doerr/AGE fotostock/firstlight.ca
pp. 2–3 Ian Crysler; p. 4 Canadian Press/Alex Galbraith; p. 7 Ray Boudreau; p. 12 Ray Boudreau; p. 15 Corel Collections, *Desserts*; p. 19 Ray Boudreau; p. 24 Ray Boudreau; p. 26 © Royalty-Free/CORBIS/MAGMA; p. 27 John A. Rizzo/Photodisc/Getty Images; p. 29 Ken Straiton/firstlight.ca; p. 31 Ian Crysler; p. 32 Ian Crysler; p. 34 (top) AP Photo/Jozsef Balaton/mti; p. 34 (bottom) Ian Crysler; p. 35 Ian Crysler; p. 37 Ian Crysler; p. 40 Karl Weatherly/Photodisc/Getty Images; p. 41 Ian Crysler; p. 42 Canadian Press/AP Photo/Francois Mori; p. 43 Ian Crysler; p. 49 (bottom) Rob Melnychak/Photodisc/Getty Images; p. 49 (inset) Skip Nall/Photodisc/Getty Images; p. 50 Ian Crysler; p. 51 Ian Crysler; p. 53 Corel Collections *Nesting Birds*; p. 56 Ian Crysler; p. 57 Ian Crysler; p. 60 Ian Crysler; p. 61 Ian Crysler; p. 63 (top) Mitch Diamond/Index Stock Imagery; p. 63 (bottom) Ian Crysler; p. 64 (top) Digital Vision/firstlight.ca; p. 64 (bottom) Canadian Press/AP Photo/Ann. M. Job; p. 67 Ian Crysler; p. 68 Ian Crysler; p. 70 Digital Vision/Getty Images; p. 71 Ian Crysler; p. 72 Corel Collection *Recreational Activities*; p. 76 (top) Larsh Bristol Photography; p. 76 (middle) Corel Collection *Barns and Farms*; p. 76 (bottom) Ian Crysler; p. 77 (top) S. Meltzer/Photolink/Getty Images; p. 77 (bottom) Ian Crysler; pp. 78–79 Corel Collection *Bridges*; p. 80 Ian Crysler; p. 84 Ian Crysler; p. 87 Ian Crysler; p. 90 Used by permission of Tangrams: www.tangrams.ca, Vancouver, Canada; p. 94 Ray Boudreau; p. 96 Ian Crysler; p. 99 Ian Crysler; p. 106 Ian Crysler; p. 108 Ian Crysler; p. 109 Ian Crysler; p. 110 Dave Starrett; p. 111 Royal Canadian Mint; p. 115 CP Photo/*Winnipeg Free Press*/Ken Gigliotto; p. 116 Ray Boudreau; p. 118 Canadian Press/AP Photo/David J. Phillips; p. 119 Canadian Press/AP Photo/Douglas C. Pizac; p. 127 Ray Boudreau; p. 131 Ian Crysler; p. 132 Ian Crysler; p. 133 Canadian Press/Robert Dall; p. 137 Ray Boudreau; p. 143 Ray Boudreau; p. 144 (top) Ray Boudreau; p. 144 (bottom) Corel Collections *Canada*; p. 152 Ian Crysler; p. 152 (inset left) Michael Freeman/CORBIS/MAGMA; p. 152 (inset right) TPL Distribution Limited/firstlight.ca; p. 154 Ian Crysler; p. 156 Canadian Press Jacques Boissinot; p. 157 Ray Boudreau; p. 158 Ian Crysler; p. 163 © Royalty-Free/CORBIS/MAGMA; p. 167 Comstock Images www.comstock.com; p. 169 Graham Nedin/CORBIS/MAGMA; p. 171 Photodisc/Getty Images; p. 172 Ray Boudreau; p. 174 Comstock Images www.comstock.com; p. 175 Ian Crysler; p. 176 (top) Canadian Press/Kevork Djansezian; p. 176 (bottom) Canadian Press/Johnathan Hayward; p. 178 Corel Collection *Ski Scenes*; p. 182 Ian Crysler; p. 184 Canadian Press/AP Photo/ Kevork Djansezian; p. 189 © Bettmann/CORBIS/MAGMA; p. 190 Ray Boudreau; p. 195 Ken Straiton/firstlight.ca; p. 197 Ray Boudreau; p. 199 Jeff Greenberg/PhotoEdit Inc.; p. 202 Ray Boudreau; p. 205 Ian Crysler; p. 208 Ray Boudreau; p. 210 Ray Boudreau; p. 213 Ray Boudreau; p. 214 Ray Boudreau; p. 216 Ray Boudreau; p. 219 (top) Ray Boudreau; p. 219 (bottom) Corel Collection *Zimbabwe*; p. 220 (top) Photo Researchers Inc.; p. 220 (bottom) Norbert Rosing/National Geographic Image Collection; p. 221 Ray Boudreau; p. 225 © Horace Bristol/CORBIS/MAGMA; p. 226 (top) M.C. Escher's Butterflies © 2004 The M.C. Escher Company – Baarn, Holland. All Rights Reserved; p. 226 (bottom) M.C. Escher's Seahorses © 2004 The M.C. Escher Company – Baarn, Holland. All Rights Reserved; p. 227 and background pp. 226–227 M.C. Escher's Horses © 2004 The M.C. Escher Company – Baarn, Holland. All Rights Reserved; p. 228 Ian Crysler; p. 229 Ian Crysler; p. 232 Ian Crysler; p. 233 Photodisc/Getty Images; p. 236 Ian Crysler; p. 239 Ray Boudreau; p. 240 Ian Crysler; p. 243 Ian Crysler; p. 244 Ian Crysler; p. 246 Ian Crysler; p. 249 Ian Crysler; p. 250 Gunter Marx Photography/CORBIS/MAGMA; p. 254 Ian Crysler; p. 255 M.C. Escher's Birds © 2004 The M.C. Escher Company – Baarn, Holland. All Rights Reserved; pp. 256–257 Ian Crysler; p. 258 Deborah Davis/PhotoEdit Inc.; p. 260 Ian Crysler; p. 262 Ian Crysler; p. 266 Ian Crysler; p. 271 Ian Crysler; p. 279 Ian Crysler; p. 280 Ian Crysler; p. 283 Ian Crysler; p. 284 Ian Crysler; p. 286 Ian Crysler; p. 288 Ian Crysler; p. 291 Jeremy Hoare/Life File/Getty Images; p. 294 Ian Crysler; p. 296 Courtesy of Lightwater Valley: www.lightwatervalley.co.uk/home.htm; p. 304 Ian Crysler; p. 312 Dorling Kindersley Media Library; p. 313 Ray Boudreau; p. 314 Michael Dick, Animals Animals/Earth Sciences; p. 318 Ray Boudreau; p. 320 Ray Boudreau; p. 322 Ray Boudreau; p. 325 Ray Boudreau; p. 329 Ian Crysler; p. 333 Ray Boudreau; p. 334 Ray Boudreau; p. 337 Ray Boudreau; p. 346 Nik Wheeler/CORBIS/MAGMA; p. 346 (inset) Johnathan Nourok/PhotoEdit Inc.; p. 347 David Muir/Masterfile; p. 347 (inset) AGE footstock; p. 348 Ian Crysler; p. 352 Ian Crysler; p. 356 © David Noton/Masterfile www.masterfile.com; p. 357 Zefa/Star/Masterfile www.masterfile.com; p. 359 (top) Jim McGuire/Index Stock Imagery; p. 359 (bottom) AFP/Getty Images; p. 362 Ian Crysler; p. 364 Ian Crysler; p. 369 Photodisc/Getty Images; p. 370 Ian Crysler; p. 371 Elizabeth Whiting and Associates; p. 374 Ian Crysler; p. 375 Ian Crysler; p. 376 Ray Boudreau; p. 377 Ian Crysler; p. 379 TPL Distribution Limited/firstlight.ca; p. 381 Ian Crysler; p. 383 Ian Crysler; p. 385 Ian Crysler; p. 386 Ian Crysler; p. 389 Nancy P. Alexander/PhotoEdit Inc.; p. 390 Ian Crysler; p. 391 Ian Crysler; p. 392 Eyewire (Photodisc)/Getty Images; p. 393 Ian Crysler

Illustrations

Linda Hendry
Paul McCusker
Dusan Petricic
Michel Rabagliati
Bill Slavin
Craig Terlson